Physics
in Life Sciences

K. H. Hausser   H. R. Kalbitzer

NMR in Medicine and Biology

Structure Determination, Tomography, In Vivo Spectroscopy

With 136 Figures

Springer-Verlag

Berlin Heidelberg New York
London Paris Tokyo
Hong Kong Barcelona
Budapest

Professor Dr. *Karl H. Hausser*
Priv.-Doz. Dr. Dr. *Hans R. Kalbitzer*

Max-Planck-Institut für Medizinische Forschung, Jahnstraße 29,
W-6900 Heidelberg, Fed. Rep. of Germany

Title of the original German edition: *NMR für Mediziner und Biologen*
© Springer-Verlag Berlin Heidelberg 1989

Library of Congress Cataloging-in-Publication Data, Hausser, K. H. [NMR für Mediziner und Biologen. English]. NMR in medicine and biology: structure and determination, tomography, in-vivo spectroscopy / K. H. Hausser, H. R. Kalbitzer. p. cm. Translation of: NMR für Mediziner und Biologen. Includes bibliographical references. Includes index.
1. Nuclear magnetic resonance spectroscopy. 2. Magnetic resonance imaging. I. Kalbitzer, H. R. (Hans R.), 1949–. II. Title. [DNLM: 1. Nuclear Magnetic Resonance. QT 34 H377n] QP519.9.N83H38 1991 574'.028–dc20 DNLM/DLC for Library of Congress 90-10426

ISBN-13: 978-3-642-76106-5   e-ISBN-13: 978-3-642-76104-1
DOI: 10.1007/978-3-642-76104-1

© Springer-Verlag Berlin Heidelberg 1991

Softcover reprint of the hardcover 1st edition 1989

56/3140 - 5 4 3 2 1 0 - Printed on acid-free paper

of NMR, whereas Chap. 4 deals with imaging by NMR tomography. Readers interested in the possibilities of in vivo spectroscopy will find when reading Chap. 5 that it contains references mainly to Chaps. 2 and 4.

We are much obliged to Professor Ch. Poole for critically reading the english version and for the valuable advice he has given. We are grateful to Drs. M. Deimling and H. Friedburg, and Professors U. Haeberlen and B. Stroebel for revising parts of the manuscript and for the valuable hints they gave us. We thank the irreplaceable Mrs. Thurm for typing the manuscript and likewise Mrs. P. Bele for providing the drawings. Our thanks for some of the drawings go also to Mrs. G. Eulefeld and Mrs. H. Kessel. Several colleagues have supported our efforts with detailed discussions and have contributed constructively to this book by providing NMR pictures and data. These colleagues were: Drs. W. Hartl, W. Hartung, W. Kuhn, E. Mueller, P. Neidig, H. Post, E.R. Reinhard, N. Schuff, J.S. Wray; to all of them go our sincere thanks.

Heidelberg, December 1990   *K.H. Hausser H.R. Kalbitzer*

Preface

About 10 years ago NMR Fourier spectroscopy began to show results relevant to the structure of larger molecules of biological interest; today it is the most frequently quoted method in biochemical papers. The many possible applications of NMR tomography and its convincing results mean that nowadays NMR tomography is attracting ever increasing interest within medicine. Another application of spatial selective NMR, in vivo NMR spectroscopy, is still at an early stage of technical development and requires more and greater efforts for its wider application. However, in the long run, the potential applications of in vivo spectroscopy are as important as NMR tomography: if it becomes possible to perform analytical investigations with biomolecules non-invasively in a well-defined volume element within the human body, that is to do medical laboratory tests inside the human body, then this will open the way for important new diagnostic techniques.

To understand the possible present and potential future applications of NMR in biology and medicine, it is essential for physicians and biologists to acquaint themselves with the basic principles of NMR. However, as most textbooks are written by physicists or physicochemists, they are not so suitable for physicians and biologists because of the mathematical formulae and the type of language used. There exist a number of reviews on the application of NMR tomography to specially selected parts of the body and to certain diseases, but their emphasis lies entirely on the medical side, the basic principles are usually dealt with only briefly, and spectroscopy is not included at all. On the other hand, it is comparatively easy for readers having a scientific/physical background to understand the basic physical principles, but they frequently lack knowledge of the medical and biological motivation needed to successfully apply NMR.

We believe that the various applications of NMR in biology and medicine should not be separated but should be dealt with together because of the many interconnections that exist. Hence, we have tried to write a book that is suitable for biologists and physicians, a book that does without mathematical formalism, as far as this is possible, and emphasizes the physical principles, and a book that presents typical biological and medical problems and their solutions by NMR.

The book is divided into five chapters each of which is self-contained, but a large number of references point out the interconnections between the different parts. Chapter 1 is the indispensable basis for understanding all subsequent parts; for this reason most of the references refer to Chap. 1. Chapters 2 and 3 are meant mainly for readers who are primarily interested in the biochemical applications

Foreword

Professor Karl Hausser has been active in NMR research for several decades, and he has made many significant contributions to the field. Professor H.R. Kalbitzer entered this research area more recently and has contributed a great deal to the medical and biological applications of NMR. It is fortunate for all of us that they have joined their efforts in producing this volume.

A great deal of current NMR research involves medical and biological applications which are of interest to a diverse audience, including physicians, biochemists, radiologists and biophysicists. These researchers approach the subject from different backgrounds and often concentrate their efforts on a narrow portion of the literature. A number of more specialized introductions and advanced monographs have been written on various aspects of this field to satisfy their needs, but is easy to lose an overall perspective.

With the appearance of the German edition of the present volume in 1989 Professors Hausser and Kalbitzer provided the German-speaking audience with a complete coverage of the basic principles and aspects of NMR from a unified perspective. The advantage of this approach was evident to readers of the German edition. Through the appearance of the present English edition it is now available to a much wider audience.

In addition to introducing the reader to the topic of medical and biological NMR this volume provides an overview on the present status of research in the field, and the well-chosen bibliography constitutes an entry to the specialized literature on the subject.

Charles P. Poole, Jr., Ph.D.
Professor of Physics
University of South Carolina
Columbia

larly, the solvent suppression techniques, originally developed for high-resolution NMR, were subsequently adapted for MRS studies of metabolism. Therefore, it is useful for all scientists in the field to have a general knowledge of the uses of NMR and to be aware of advances in the field even though they may not initially apply directly to one's own interests.

A word of advice to the newcomer to NMR: magnetic resonance is a phenomenon that allows detection of signals from nuclei. The NMR techniques (from magnet design, high-resolution, 2- or 3-dimensional NMR, MRI, and MRS) are fascinating tools, but they remain techniques. These methods are a "means to an end", in this case the end being the answer to the scientific question you wish to answer. Some scientists seem to become so involved in the development of NMR techniques that they lose sight of the questions that they are investigating. While technique development is important, the best techniques are developed in order to answer a well-focused scientific question.

In summary, NMR as presented in this book is an exciting technology which will provide much greater understanding of the structure and function of macromolecules, cells, and organs.

Michael W. Weiner, M.D.
Professor of Medicine
and Radiology
University of California
San Francisco

Foreword

During the past decade, magnetic resonance has emerged from being a relatively obscure measuring tool of use only to chemists and physicists and has become a widely used modality in all areas of biological science, clinical investigation, and medical diagnosis. In fact, applications of magnetic resonance have become so widespread that all scientists and physicians should have some knowledge of the fundamental principles and applications of NMR spectroscopic measurements of molecular structure, cell metabolism, and body anatomy. Thus, it is appropriate that the English translation of *NMR in Medicine and Biology* by K.H. Hausser and H.R. Kalbitzer is now available to English-speaking physicians and scientists.

Frequently I am asked by my medical colleagues, students, and scientists new to the field, "How can I learn the general principles of nuclear magnetic resonance spectroscopy and imaging?" This text provides the answer to this question. It begins with a clear introduction to the basic principles of NMR, the NMR spectrum, and the principles of relaxation. Subsequently, the reader is progressively guided through the increasingly sophisticated techniques which are used to probe macromolecular structure. Finally, general principles of magnetic resonance and in vivo spectroscopy for the study of metabolism are discussed. I have found that it is not necessary to begin at the beginning; those who are interested in particular applications of NMR can begin reading at any point. If the language appears too sophisticated, the reader can then read earlier sections in order to understand the terminology fully.

Although it is often fashionable to recall the historic development of techniques, I find it more exciting to contemplate future applications and discoveries of NMR, especially considering the rapid growth of the field in the past few years. NMR in biology has two major currents of activity: NMR spectroscopy to determine macromolecular structures and magnetic resonance imaging/spectroscopy for clinical studies of anatomy and physiology and metabolism. These two areas are bridged by a common interest in technology including magnet design, improved gradients, pulse sequences, and signal processing. A common goal of all spectroscopists is to improve signal-to-noise and spectral resolution. This is achieved through higher-field magnets, higher-sensitivity probes, optimized pulse sequences, and improved techniques for signal processing. Although there is always a tendency to be knowledgeable in one's narrow area of interest, many advances in NMR applications have often been derived from ideas developed for other applications. For example, two-dimensional NMR was originally developed for studies of chemical structure and was then adapted to NMR imaging. Simi-

Contents

List of Abbreviations

ADC	Analog-to-Digital Converter
$B_0, \mathbf{B_0}$	Static magnetic field (magnetic induction)
$B_1, \mathbf{B_1}$	Radio frequency magnetic field (magnetic induction)
CPMG	Carr-Purcell-Maiboom-Gill
CPU	Central Processing Unit
cw	Continuous wave
D	Deuteron, ^2H
e	Elementary charge (1.60219×10^{-19} A s)
FID	Free Induction Decay
G_x, G_y, G_z	Magnetic gradient fields along the axis of the cartesian coordinate system
h	Planck's constant (6.6262×10^{-34} m^2 kg s^{-1})
$\hbar$	$h/2\pi$ (1.0546×10^{-34} m^{2}kg s^{-1})
Hz	Hertz, unit of frequency
I	Nuclear spin, quantum number
J	Scalar spin-spin coupling measured in Hertz
k	Boltzmann's constant (1.3806×10^{-23} m^{2}kg s^{-2}K^{-1})
kHz	1000 Hertz
M	Magnetization; unit of concentration mol/l*
$\mathbf{M}$	Magnetization
M_x, M_y	Components of magnetization perpendicular to the magnetic field
M_z	Component of magnetization parallel to the magnetic field
M_0	Magnetization in Boltzmann equilibrium
MIPS	Million instructions per second
NMR	Nuclear magnetic resonance
N	Avogadro's number (6.02217×10^{23}mol^{-1})
P_i	Inorganic phosphate
pixel	Picture element
ppm	Parts per million, $1 : 10^6$
$Q \cdot e$	Nuclear quadrupole moment
$\mathbf{r}$	Radius vector in the spherical coordinate system
rf	Radio frequency
S	Electronic spin
s	Saturation parameter
S/N	Signal/noise

T	Absolute temperature; Tesla, unit of magnetic induction*
T_1	Spin-lattice or longitudinal relaxation time
T_2	Spin-spin or transversal relaxation time
T_{2*}	Effective transversal relaxation time
T_E	Spin echo time between 90° pulse and spin echo
T_R	Repetition time of subsequent pulse sequences
voxel	Volume element, corresponds in 3D space to a pixel in 2D space
α	Angle, flip angle
γ	Magnetogyric ratio
δ	Chemical shift measured in ppm
Θ	Azimuth angle in the spherical coordinate system
μ_I	Magnetic moment of the nuclear spin I
μ_S	Magnetic moment of the electronic spin S
μ_0	Magnetic permeability of the vacuum $(4\pi \times 10^{-7}\,\mathrm{m\,kg\,s^{-2}A^{-2}})$
ν	Frequency measured in Hertz
ϱ	Density; spin-lattice relaxation rate*
σ	Shielding constant; cross relaxation rate*
τ	Correlation time
τ_{rot}	Rotational correlation time
τ_{trans}	Translational correlation time
τ_S	Correlation time for paramagnetic relaxation
τ_{sc}	Correlation time for scalar interaction
φ	Angle in the spherical coordinate system; dihedral angle in the peptide main chain*
χ	Dihedral angle in the peptide side chains; magnetic susceptibility*
ψ	Dihedral angle in the peptide main chain
ω	Angular frequency $(2\pi\nu)$; dihedral angle in the peptide main chain*

---

* In some cases the same symbol is used for different terms.

1. Principles of NMR

The purpose of Chap. 1 of this book is to lay the foundation for a general under-standing of NMR, which is then built upon in the following chapters. Nuclear magnetic resonance is just one of the names of the method treated in this book; other terms used are nuclear spin resonance or the abbreviation NMR.

The first researchers to succeed in detecting a NMR signal were *Felix Bloch* (Bloch 1946) and, independently, *Edward Purcell* (Purcell et al., 1946). The next important step was the discovery of the chemical shift, that is a variation of the NMR frequency caused by the electronic cloud that surrounds the nucleus in metals (Knight, 1949) and in liquids (Arnold et al., 1951). The existence of the chemical shift is one of the main effects that forms the basis for the many applications of NMR in other fields besides physics. It allows one to draw conclusions concerning the distribution of electrons within the molecule and thus the chemical structure. Figure 1.1 shows the proton resonance spectrum of ethanol reproduced from the original paper by Arnold et al. (1951). This spectrum has opened a new field of research, which is today known as "high resolution NMR spectroscopy in liquids". Nowadays the majority of all NMR experiments in chemistry, biology and medicine are performed in this field. Even imaging by NMR, NMR tomography, is basically NMR in liquids. However, in NMR tomography the chemical shift is usually a disturbing factor that is to be suppressed by adequate methods. For physical reasons high resolution NMR can only be performed in the liquid state, but by using special experimental methods it is also possible to obtain important information from NMR in the solid state. A drawback of NMR is its comparatively low sensitivity, but this was greatly

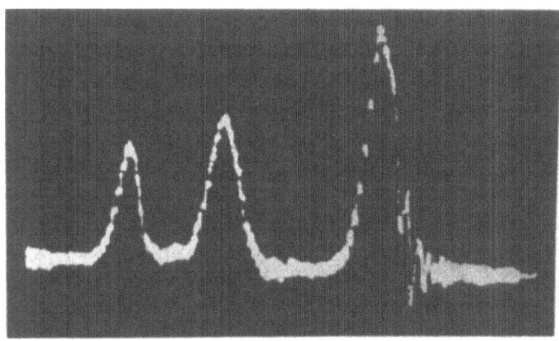

Fig. 1.1. ^{1}H-NMR spectrum of ethanol (Arnold et al., 1951, with permission), first high resolution NMR spectrum of a liquid as published. The signals can be assigned from left to right to the protons of the hydroxyl, methylene and methyl groups of ethanol

improved by the Fourier spectroscopy introduced by *Richard Ernst* (Ernst and Anderson, 1966) and the development of superconducting high-field magnets.

A milestone for the investigation of biological macromolecules was the introduction of the two-dimensional NMR spectroscopy (2D NMR) first proposed by Jean Jeener (Jeener, 1971). 2D NMR is still a field of active research in physics and novel 2D NMR experiments are regularly published. In the biologically oriented fields of research 2D NMR is the basis for determining the structure of proteins in solution and is now in the process of becoming an alternative to the X-ray structure analysis of small proteins. Part of the pioneering work in this connection was done by *Kurt Wüthrich* and his group.

Space selection using a field gradient was independently proposed by *Peter Mansfield* (Mansfield, 1973) and by *Paul Lauterbur* (Lauterbur, 1973). The latter showed that a suitable space selection can lead to a novel imaging method. However, it is to be expected that spectroscopic methods will also gain in importance in medicine.

1.1 The Basic Principle of Magnetic Resonance

Processes at the elementary level of single particles and molecules can in general only be completely described using quantum mechanics. Quantum mechanics is an abstract theory at a high mathematical level and its contents are expressed through the relations between operators. These operators do not have a direct physical significance, but their expectation values, that is the values obtained when measuring a sufficiently large number of particles, have. The calculation of these expectation values is given by the theory and is symbolized by the pair of brackets $\langle \rangle$. For instance, the energy E is given by the expectation value of the Hamiltonian $\mathcal{H}$ of the system, that is $E = \langle \mathcal{H} \rangle$. If the brackets are omitted, the equations between operators and those between the corresponding expectation values appear in many cases to be formally equal although their meaning is quite different. We shall in general restrict ourselves to the expectation values and use the bracket symbol only when necessary in cases of doubt.

In a practical NMR experiment the signals originate from a macroscopic sample, which consists of a large number of molecules. One of the main criteria for the correctness of microscopic theories like quantum mechanics is that, when passing to the macroscopic level, they must give the same results as the well-established theories. The basic principles of magnetic resonance can therefore be understood within the framework of classical physics if just a few additional assumptions are made that reflect the properties of quantum mechanics. Because of its simplicity and clearness we shall use this semi-classical description of magnetic resonance as much as possible.

1.1.1 Magnetic Moment and Nuclear Spin

Most nuclei possess a spin angular momentum J and can be regarded as microscopically small spinning tops. The angular momentum J is always connected with a magnetic moment μ_I, and between them there exists the simple relation

$$\mu_I = \gamma_I J = \hbar \gamma_I I \quad . \tag{1.1}$$

The proportionality constant γ_I is termed magnetogyric ratio; it is a constant that is characteristic for each type of nucleus. In most cases the nuclear spin J is expressed in terms of Planck's constant $\hbar(\hbar = h/2\pi)$ and is described by the dimensionless symbol I. The magnitude of the nuclear spin I is also an invariant property of each nucleus (more precisely, of each nucleus in the ground state, because in high energy physics excited states with a different nuclear spin are quite common); it is usually characterized by the spin quantum number I, which for theoretical reasons can only assume multiple values of 1/2. In quantum mechanics the absolute value of the angular momentum J is given by

$$\langle |J| \rangle = \hbar \langle |I| \rangle = \hbar \sqrt{I(I+1)} \quad . \tag{1.2}$$

The larger γ_I and I are, the larger is the magnetic moment of the nucleus, that is, the larger is the magnetic field of our microscopic magnet.

The nucleus of the hydrogen atom ^1H possesses the largest magnetogyric ratio of all stable isotopes. Since the sensitivity of detection increases with the magnetic moment and, furthermore, since compounds containing hydrogen form the major part of biological systems (the human body consists of about 60 % water), ^1H-NMR is particularly important for biology and medicine. In contrast to this, the main isotopes of oxygen and carbon, ^{16}O and ^{12}C, respectively, possess a nuclear spin of $I = 0$ and are therefore unsuitable for NMR.

Other nuclei important for NMR in biochemistry and medicine are phosphorus ^{31}P and the rare isotopes of carbon ^{13}C and of nitrogen ^{15}N, because they have spin $I = 1/2$, as does the hydrogen atom ^1H. However, the natural abundance of the ^{13}C isotope is only about 1 % compared to 99 % of the main isotope ^{12}C with spin $I = 0$, and that of the ^{15}N isotope is as low as 0.4 %. Nuclei such as ^{14}N or ^{23}Na, which possess a spin $I \geq 1$ and therefore an electrical quadrupole moment, are less suitable for high resolution NMR spectroscopy. An exception to this is the heavy isotope ^2H, the quadrupolar moment of which is so small that it may be used for spectroscopic applications. Table 1.1 shows the nuclei that are most important for biological and medical applications.

1.1.2 Resonance Condition

What in fact *is* the phenomenon of nuclear magnetic resonance? If one observes a mechanical top in the gravitational field of the earth, it undergoes, under the influence of this field, a rotation around the direction of the field known as Larmor precession. Since we are always in the gravitational field of the earth, we do not usually realize that the frequency of this precession is proportional to the magnitude of the field. Hence, for example, on the moon, where the gravitational field is six times weaker than on the earth, the precession frequency is lower by a factor of 6. Analogously, a magnetic top, for instance an atomic nucleus with spin $I = 1/2$, magnetic moment μ_I and magnetogyric ratio γ_I undergoes

Table 1.1. Properties of important atomic nuclei

Istope	Nuclear spin I	Magnetogyric ratio γ_I [$T^{-1}s^{-1}$]	Resonance frequency at 14.092 T [MHz]	Natural abundance [%]	Relative sensitivity [%]
^{1}H	1/2	2.6752×10^8	600.0	99.985	100.00
^{2}H	1	4.1065×10^7	92.1	0.015	0.96
^{3}H	1/2	2.8535×10^8	640.0	–	121.36
^{12}C	0	–	–	98.89	–
^{13}C	1/2	6.7266×10^7	150.9	1.11	1.59
^{14}N	1	1.9325×10^7	43.3	99.63	0.10
^{15}N	1/2	-2.7108×10^7	60.8	0.37	0.10
^{16}O	0	–	–	99.76	–
^{17}O	5/2	-3.6267×10^7	81.4	0.04	2.91
^{18}O	0	–	–	0.20	–
^{19}F	1/2	2.5167×10^8	564.5	100.00	83.34
^{23}Na	3/2	7.0762×10^7	158.7	100.00	9.25
^{24}Mg	0	–	–	78.99	–
^{25}Mg	5/2	-1.6371×10^7	36.7	10.00	0.27
^{26}Mg	0	–	–	11.01	–
^{31}P	1/2	1.0829×10^8	242.9	100.00	6.63
^{32}S	0	–	–	95.00	–
^{33}S	3/2	2.0518×10^7	46.0	0.76	0.23
^{34}S	0	–	–	4.22	–
^{35}Cl	3/2	2.6213×10^7	58.8	75.77	0.47
^{37}Cl	3/2	2.1819×10^7	48.9	24.23	0.27
^{39}K	3/2	1.2484×10^7	28.0	93.26	0.05
^{41}K	3/2	6.8521×10^6	15.4	6.73	0.01
^{40}Ca	0	–	–	96.94	–
^{43}Ca	7/2	-1.8000×10^7	40.4	0.14	0.64
^{110}Cd	0	–	–	12.40	–
^{111}Cd	1/2	-5.6729×10^7	127.2	12.86	0.95
^{112}Cd	0	–	–	24.00	–
^{113}Cd	1/2	-5.9344×10^7	133.1	12.34	1.09
^{114}Cd	0	–	–	28.70	–
^{116}Cd	0	–	–	7.60	–

The NMR detectability in a given magnetic field is to a first approximation proportional to $\gamma_I^3 I(I+1)$ if the relaxation times are equal for all nuclei. As we shall see later, this is a strong simplification that is in general not valid under practical conditions. The sensitivity of detection of the ^{1}H nucleus was defined to be 100 % as usual.

in an external magnetic field B_0[1] a Larmor precession with angular frequency $\omega_I = 2\pi\nu_I$. Quantitatively, the larger the magnetogyric ratio γ_I of the nucleus and the larger the magnetic field B_0, the higher is this angular frequency, that is, $\omega_I = \gamma_I B_0$.

If an alternating magnetic field B_1, which might be produced by a radio frequency coil, in addition to the static magnetic field B_0, is applied to the nuclear spins I, in general nothing will happen. However, if the angular frequency ω of this alternating magnetic field equals the Larmor frequency of the atomic nuclei ω_I, one observes a very strong interaction. This "resonance effect" is widespread in our physical world: for instance, a swing can be made to reach its greatest height if it is always pushed at the very same moment of each swing cycle.

The resonance condition expressed by the simple mathematical relation

$$\omega = \omega_I = \gamma_I B_0 \tag{1.3}$$

is the basic equation of NMR.

Let us now consider the principle of NMR from the viewpoint of quantum mechanics. Without an external magnetic field the many small magnetic dipoles μ_I, for instance of hydrogen atoms in a glass of water, are oriented at random. If a static magnetic field B_0 is applied to this system of nuclear spins, then these small elementary magnets orient themselves with respect to this field. Following a fundamental law of physics first discovered by Stern and Gerlach in 1923, such particles with spin $I = 1/2$ cannot orient themselves at random with respect to the external magnetic field B_0 but rather do so either "parallel" or "antiparallel" to the field corresponding to the two magnetic quantum numbers $m_I = +1/2$ and $-1/2$. This law is known as spatial quantization.

The Cartesian coordinate system in NMR is usually chosen in such a way that the z-axis is parallel to the external magnetic field B_0. The expectation value of the z-component I_z of the spin I is then given by

$$\langle I_z \rangle = \hbar m_I \quad . \tag{1.4}$$

The energy E_m of a magnetic dipole μ in an external magnetic field B_0 is given by

$$E_m = -\mu B_0 = -\hbar\gamma_I m_I B_0 \quad . \tag{1.5}$$

Here, μ is expressed in terms of the nuclear spin (1.1). For the particle with spin $1/2$ in an external magnetic field B_0 one obtains two discrete energy levels E_1 and E_2 (Fig. 1.2):

$$E_1 = -\hbar\gamma_I \tfrac{1}{2} B_0 \tag{1.6}$$

$$E_2 = \hbar\gamma_I \tfrac{1}{2} B_0 \quad . \tag{1.7}$$

---

[1] The more correct but seldom used term for B_0 is magnetic flux density or magnetic induction.

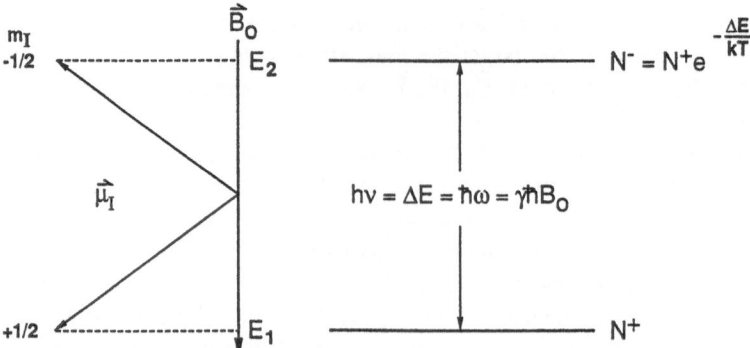

Fig. 1.2. The splitting of the nuclear spin energy levels in the static magnetic field for the case $I = 1/2$

As with any spectroscopy, transitions may be induced between the energy levels if the energy of the electromagnetic quanta $\hbar\omega$ is equal to the energy difference ΔE, that is, if the frequency of the radio frequency field (rf field) B_1 satisfies the condition

$$\hbar\omega = \Delta E = \hbar\gamma_I B_0 \qquad (1.8)$$

Division by $\hbar$ yields the basic equation of NMR (1.3), which was already derived from intuitive arguments.

1.1.3 Bloch Equations

The number of atomic nuclei in a macroscopic sample is very large. One cubic centimeter of water contains several 10^{22} ^{1}H nuclei, the small magnetic dipoles of which are oriented at random. In a *Gedankenexperiment* we shall now apply a static magnetic field B_0 in a very short time. Then the ^{1}H spins with $I = 1/2$ will be statistically oriented with respect to the B_0-field, that is, with half of them parallel and the other half antiparallel. However, in this state our nuclear spin system is not in thermal equilibrium with its surroundings. If we denote with N^+ the number of spins I that have z-components I_z oriented parallel to B_0, and with N^- the number of spins that have I_z oriented antiparallel, the ratio of N^- to N^+ in thermal equilibrium with the surroundings is given by the Boltzmann factor:

$$\frac{N^-}{N^+} = e^{-\Delta E/kT} \qquad (1.9)$$

Here, k is Boltzmann's constant.

The lower energy level is somewhat more populated, that is more magnetic moments are oriented parallel to the external magnetic field than antiparallel. At room temperature and with the usual laboratory magnetic fields of the order of $1\,\mathrm{T}$ ($= 10000\,\mathrm{G}$) the difference in population is about 10^{-6}. This small surplus of spins oriented parallel to the magnetic field – there are still of the order of

10^{16} spins per cm^3 of matter – suffices to produce a measurable macroscopic magnetization M.

In thermal equilibrium the resulting macroscopic magnetization M is parallel to the external magnetic field B_0. The magnitude of the magnetization M_0 at room temperature (the high temperature approximation) can be derived from (1.9) to be

$$M_0 = \frac{N\hbar^2\gamma_I^2 I(I+1)}{3kT}B_0 \quad , \tag{1.10}$$

where $N = N^+ + N^-$ is the total number of nuclear spins. As can be seen, the macroscopic magnetization increases with increasing magnetic field B_0, with increasing magnetogyric ratio γ_I and with decreasing temperature T. Many effects occurring in NMR are determined by the behavior of the magnetization M.

The process by which temperature equilibrium is reached between the spin system and its surroundings, which in general and even in liquids is called "lattice", is termed "spin-lattice relaxation". It describes the process by which the z-component of the magnetization M_z reaches its equilibrium value M_0 after a perturbation, if one waits long enough. Since the equilibrium magnetization is oriented parallel to the external magnetic field B_0, the spin-lattice relaxation is also termed "longitudinal relaxation".

Soon after the discovery of NMR, Felix Bloch classically described the behaviour of the magnetization M with its components M_x, M_y and M_z by a system of differential equations, the Bloch equations. They describe in a simple way many important experimental results: (1) If magnetization and magnetic field do not have the same orientation, the magnetization precesses around the magnetic field. (2) If one waits long enough after a perturbation, equilibrium magnetization is reached; the M_z component parallel to the magnetic field equals M_0 and the transverse components perpendicular to B_0 vanish. The exponential approach of M_z to the value of M_0 in Boltzmann equilibrium is described by the following equation:

$$\frac{dM_z}{dt} = \frac{M_0 - M_z}{T_1} \quad . \tag{1.11}$$

The time constant T_1 is called the "longitudinal relaxation time". Accordingly, the decay of the transverse components M_x and M_y is described by

$$\frac{dM_x}{dt} = -\frac{M_x}{T_2} \quad , \quad \frac{dM_y}{dt} = -\frac{M_y}{T_2} \quad . \tag{1.12}$$

By combining (1.11) and (1.12) with the classical equation of motion for the precession of the magnetization

$$\frac{dM}{dt} = -\gamma_I(B \times M) \quad , \tag{1.13}$$

one obtains the Bloch equations

$$\frac{dM_x}{dt} = -\gamma_I (\boldsymbol{B} \times \boldsymbol{M})_x - \frac{M_x}{T_2} \qquad (1.14)$$

$$\frac{dM_y}{dt} = -\gamma_I (\boldsymbol{B} \times \boldsymbol{M})_y - \frac{M_y}{T_2} \qquad (1.15)$$

$$\frac{dM_z}{dt} = -\gamma_I (\boldsymbol{B} \times \boldsymbol{M})_z + \frac{M_0 - M_z}{T_1} \quad . \qquad (1.16)$$

In a typical NMR experiment, in addition to the static magnetic field B_0 in the z-direction, an rf field is applied that is polarized perpendicular to the magnetic field B_0, for instance in the x-direction, and that oscillates with frequency ω. This field $2B_1 \cos \omega t$ is in general much weaker than the external magnetic field B_0. A linearly polarized alternating field can be divided into two circularly polarized components, which then rotate with an angular frequency $\pm\omega$, one clockwise and the other anticlockwise. While the interaction of the nuclear spins with the component rotating in the opposite direction to the Larmor precession is negligibly small, the other component is "seen" by the nuclear spins rotating in the same direction as a static magnetic field B. Hence, the magnetic field B that affects the magnetization M is

$$\boldsymbol{B} = B_1 \cos \omega t \, \boldsymbol{i} + B_1 \sin \omega t \, \boldsymbol{j} + B_0 \, \boldsymbol{k} \quad , \qquad (1.17)$$

where $\boldsymbol{i}$, $\boldsymbol{j}$ and $\boldsymbol{k}$ are unit vectors in the directions x, y and z. If they are inserted into the Bloch equations, we obtain

$$\frac{dM_x}{dt} = \gamma_I (M_y B_0 - M_z B_1 \sin \omega t) - \frac{M_x}{T_2} \qquad (1.18)$$

$$\frac{dM_y}{dt} = \gamma_I (M_z B_1 \cos \omega t - M_x B_0) - \frac{M_y}{T_2} \qquad (1.19)$$

$$\frac{dM_z}{dt} = \gamma_I (M_x B_1 \sin \omega t - M_y B_1 \cos \omega t) + \frac{M_0 - M_z}{T_1} \quad . \qquad (1.20)$$

It has proved useful in many cases to describe the behavior of the magnetization M in a coordinate system $(x', y', z' = z)$ that rotates with angular frequency ω around the z-axis. With this coordinate transformation it is possible to formally simplify the equations, and, moreover, it is illustrative as well. The complicated motion of the magnetization in space can be divided into two individual motions: the motion in the coordinate system and, superposed onto it, the rotation of this coordinate system with respect to the static laboratory system. Quite often the rotation frequency ω_r of the coordinate system is chosen to be equal to the radio frequency ω, since then the B_1-field in the rotating system becomes static. If we denote by $M_{x'}$ the in-phase component of the magnetization parallel to the x'-axis, that is parallel to B_1, and $M_{y'}$ the component parallel to the y'-axis with a phase shift of 90°, we obtain $(\omega = \omega_r)$

$$M_{x'} = M_x \cos \omega t + M_y \sin \omega t \qquad (1.21)$$

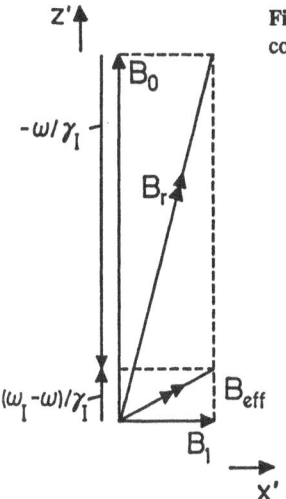

z′

$-\omega/\gamma_1$

B_0

B_r

$(\omega_I-\omega)/\gamma_1$

B_{eff}

B_1

x′

Fig. 1.3. Magnetic fields in the laboratory system and in the rotating coordinate system

$$M_{y'} = -M_x \sin \omega t + M_y \cos \omega t \quad . \tag{1.22}$$

In the rotating coordinate system the Bloch equations assume the simple form

$$\frac{dM_{x'}}{dt} = (\omega_I - \omega)M_{y'} - \frac{M_{x'}}{T_2} \tag{1.23}$$

$$\frac{dM_{y'}}{dt} = -(\omega_I - \omega)M_{x'} + \gamma_I B_1 M_z - \frac{M_{y'}}{T_2} \tag{1.24}$$

$$\frac{dM_z}{dt} = -\gamma_I B_1 M_{y'} + \frac{M_0 - M_z}{T_1} \quad . \tag{1.25}$$

Figure 1.3 gives a simple picture of the process in the rotating coordinate system. Since our coordinate system (x', y', z') rotates with the angular frequency ω of the B_1-field, B_1 is static in this coordinate system. It is useful to define, in addition to B_r, the net magnetic field resulting from B_0 and B_1, an effective magnetic field B_{eff}, which is the sum of the vectors of $(\omega_I-\omega)/\gamma_I$ and B_1. It can be shown that (1.23–1.25) describe a damped precession of the magnetization around this effective magnetic field. In the simplest case when $\omega = \omega_I$ this effective field equals B_1 and is situated in the x', y'-plane. The frequency ω_I' of the precession is given by a relation that is formally analogous to the resonance condition, but is determined by B_1 instead of B_0.

$$\omega_I' = \gamma_I B_1 \quad . \tag{1.26}$$

If the magnetization initially points in the direction of the z-axis, and if the radio frequency field is switched on for a short time t, the magnetization is turned first to the direction of the y-axis, then further to the negative z-axis, until it finally points again along the z-axis, and so on. If the B_1-field is switched off

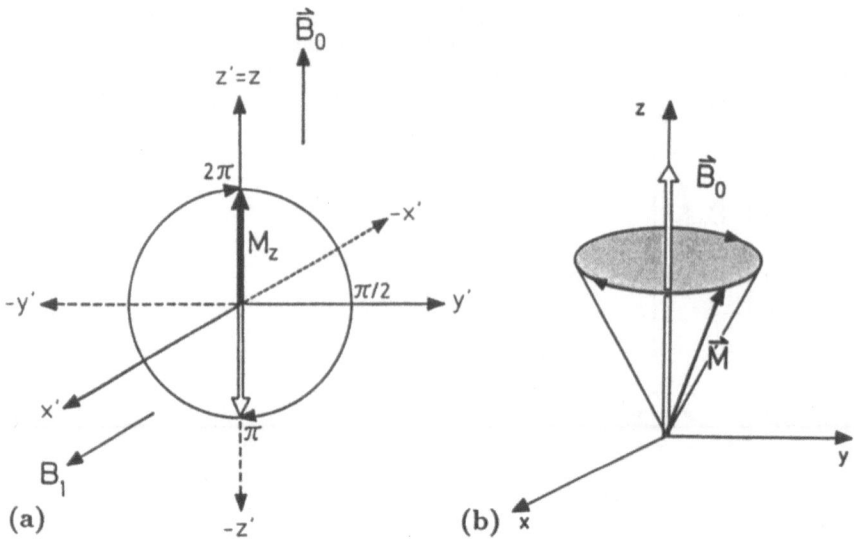

Fig. 1.4a,b. Precession motion of the magnetization. (a) Precession of the magnetization around the B_1-field in the coordinate system that rotates with the angular frequency $\omega = \omega_I$ (b) Precession of the magnetization M in the laboratory system after switching off the radio frequency field B_1. The direction of the precession shown is valid for nuclear spins with positive magnetogyric ratio

the moment the magnetization is turned by 90°, this is then termed a $\pi/2$ rf pulse or a 90° pulse. Accordingly a pulse of double length at the same B_1-field strength produces a turn of 180°, that is, a π rf pulse or a 180° pulse (Fig. 1.4a). In general, by choosing an appropriate B_{eff} and duration of the pulse t, it is possible to turn the magnetization to any desired direction.

If one diverts the magnetization from its equilibrium parallel to the z-axis with a short radio frequency pulse, it will precess around the B_0-field when the radio frequency is switched off (Fig. 1.4b). The precession of the magnetization causes a modulation of the magnetic field that is connected with the magnetization. If the sample is surrounded by a detection coil the time-dependent magnetic field induces a small voltage, which is detectable by suitable methods. The amplitude of this signal, which is termed free induction decay (FID), is proportional to the resonance frequency ω_I and to the magnetization M_0. It decays with time as the magnetization approaches its equilibrium value according to the Bloch equations.

1.1.4 Spin-Lattice Relaxation

The Bloch equations (Bloch, 1946) describe quantitatively the time development of the magnetization M_z. The solution of these equations for M_z is an exponential function with a time constant T_1, which is called the "longitudinal" or "spin-lattice relaxation time".

$$M_z(t) = M_0 - \{M_0 - M_z(0)\}\, e^{-t/T_1} \quad . \tag{1.27}$$

If the system is left to itself after a 180° pulse, it again approaches Boltzmann

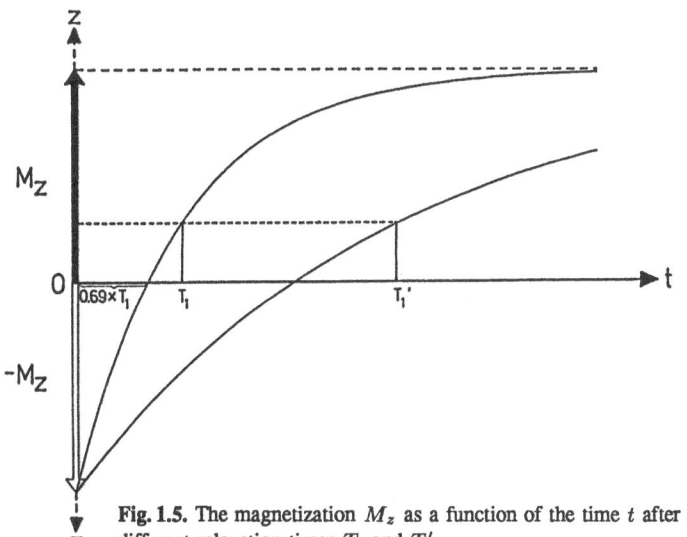

equilibrium. In particular the magnetization M_z is equal to zero in accordance with the exponential function (1.27) at $T_1 \ln 2 = 0.69 T_1$, so that T_1 can be measured by determining this zero crossing (Fig. 1.5).

When applying the usual detection system neither a negative magnetization $-M_z$ in the $(-z)$-direction nor a positive magnetization $+M_z$ will give a signal. A detection signal only originates from the tranverse component of the magnetization in the xy-plane. To determine the longitudinal relaxation time T_1, the magnetization is first flipped into the $(-z)$-direction by a 180° pulse. After waiting a certain time τ, during which the magnetization reaches the value $M_z(\tau)$ because of the longitudinal relaxation, the quantity $M_z(\tau)$ can be determined by flipping the z-magnetization with a 90° pulse into the xy-plane, where it produces an NMR signal that is proportional to $M_z(\tau)$. Since the magnetization is first inverted and later the development of the equilibrium magnetization is observed, this method is also called "inversion recovery" and symbolized as $(180°\text{-}\tau\text{-}90°)$.

1.1.5 Decay of the Transverse Magnetization and Spin-Spin Relaxation

The second type of relaxation to be considered is transverse relaxation. On the one hand, it is closely related to the line width and therefore determines the resolution of high resolution NMR spectroscopy. On the other hand, differences in the transverse relaxation times of different tissues, as well as differences in the longitudinal relaxation times, are very important for the contrast resolution of the images in NMR tomography.

If a $\pi/2$ or 90° pulse is applied to the nuclear spin system instead of a 180° pulse, the magnetization M is turned only to the direction of the y-axis. If it is then left to itself, it rotates with a Larmor precession ω_I around the magnetic field

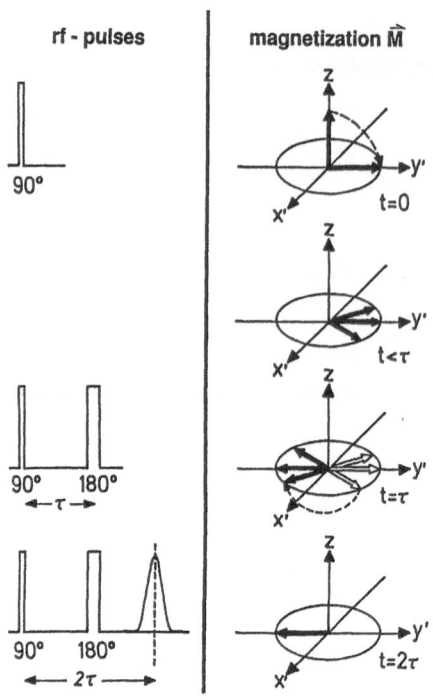

rf - pulses	magnetization $\tilde{M}$

90°

90° 180°
←— τ —→

90° 180°
←—— 2τ ——→

t=0

t<τ

t=τ

t=2τ

Fig. 1.6. Development of the magnetization with the spin echo pulse sequence. The FID produced by the 90° pulse is not shown. (*Left*) Schematic plot of the high frequency pulses and of the spin echo signals. In reality the pulses are more intense than the NMR signal by several orders of magnitude. (*Right*) Fanning out and refocusing of the magnetization in the inhomogeneous static magnetic field B_0 plotted in the rotating coordinate system

B_0 and is termed M_{xy}. However, since this B_0-field is never homogeneous over the volume of the sample, not all spins that altogether form the magnetization M_{xy} rotate with the same speed in the xy-plane: some rotate a little faster and some others a little slower than the average value. Figure 1.6 shows what happens in this case. The magnetization M_{xy} spreads out in the xy-plane until it finally points in all directions, its average becomes zero, and it therefore does not produce a detectable signal. A group of spins that rotate with the same angular frequency is usually called a "spin packet". If a 180° pulse is applied to this spin system after a time τ, all spin packets change their positions (= phase) in such a manner that the fastest are now behind and the slowest in front. What happens now can be compared with a group of runners on a track who have all started at the same time. After a certain time they are spread over the field in such a way that the faster runners are farther away from the start than the slower ones. If now the order "return" is given, the faster runners will be behind and if they continue to run at the same speed, they will have caught up with the slower ones precisely at the start and all will arrive there at the same time. The spin packets' behavior would be quite analogous if the direction of the magnetic field could be changed by 180° in a very short time. This, however, is technically impossible. A similar effect can also be achieved by a 180° pulse that changes the positions of the spin packets in such a way that the faster spin packets catch up with the slower ones after a time 2τ. Thus a resulting magnetization is again formed in the xy-plane and it produces a detection signal that was named by its discoverer Erwin Hahn the "spin echo". The resulting echo has the opposite sign with respect to the FID after the 90° pulse, it is out of phase by 180°. This can be avoided by shifting the phase of the 180° pulse by 90°, that is, in the coordinate system rotating with the frequency ω the direction of the B_1 field of the 90° pulse is oriented parallel to the x-axis and that of the 180° pulse is oriented parallel to the y-axis. The rotation around the y-axis in this case influences only those spin packets that precess faster or slower than ω_I, that is, those which dephase because of the inhomogeneity of the magnetic field B_0.

However, this is not all. In addition to this reversible dephasing process caused by the inhomogeneity of the magnetic field B_0, there also exists an irreversible dephasing process due to the interaction of the spins with one another. While the dephasing process caused by the inhomogeneity of the magnetic field is a function of the actual distribution of the B_0-field, the decay of the magnetization in the xy-plane, owing to the irreversible processes, occurs exponentially with $\exp(-t/T_2)$ following the Bloch equations. The transverse relaxation time T_2 is also known as the "spin-spin relaxation time" because it originates from the interaction of the spins with one another. In a spin echo experiment the spin-spin relaxation leads to a decay of the amplitude of the echo following the exponential function defined by the time T_2. If the amplitude of the echo is observed as a function of time between the 90° and the 180° pulse, T_2 may also be determined in inhomogeneous magnetic fields.

The determination of T_2 by this spin echo method is a somewhat clumsy procedure because many different individual spin echoes at different echo times τ have to be measured (Fig. 1.7). For this reason another pulse sequence was developed by Carr and Purcell (1954) and improved by Meiboom and Gill (1958).

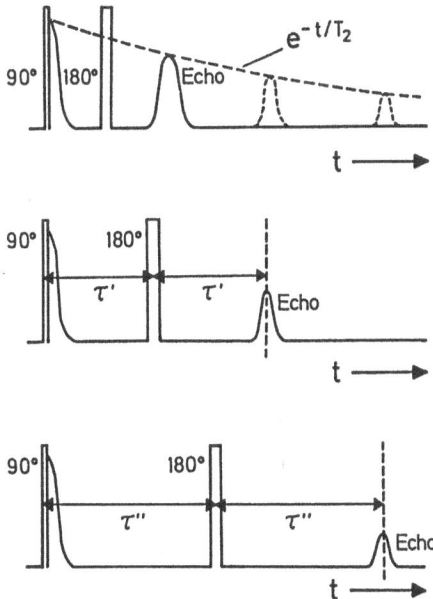

Fig. 1.7. Decrease of the spin echo amplitude as a function of the delay time τ. The decrease of the spin echo amplitude is determined by the spin-spin relaxation time T_2. In contrast to this, the building up and the decay of the individual spin echoes and of the first FID are essentially determined by the inhomogeneity of the magnetic field

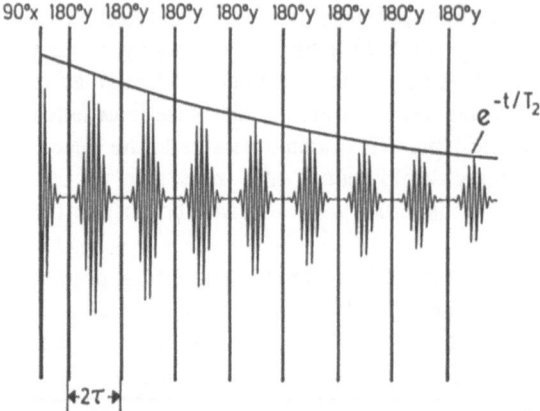

Fig. 1.8. The course of the signal with the Carr-Purcell-Maiboom-Gill (CPMG) spin echo pulse sequence

In this sequence a series of 180° pulses, during which the spin echo is constantly measured, follows a 90° pulse (Fig. 1.8). The main advantage of the CPMG spin echo sequence is that the total T_2 measurement can be achieved with one single pulse sequence. Furthermore, it allows a more precise measurement of T_2 since the precision of measurement with the simple spin echo sequence may be influenced by diffusion processes. The CPMG spin echo sequence has today gained a special importance for NMR tomography which will be demonstrated in Chap. 4.

1.2 The NMR Spectrum

By knowing the resonance condition and the definition of longitudinal and transverse relaxation, we possess the information needed to understand the principle of nuclear spin resonance experiments. This knowledge suffices to describe nuclear spin tomography, the spatial resolution of which is based on the variation of the resonance frequency resulting from space-dependent magnetic fields, and the contrast resolution of which is essentially determined by differences in the relaxation times. It is also sufficient for understanding NMR as a method of elementary analysis, since different elements and isotopes can be analysed on the basis of their different resonance frequencies. However, this information does not suffice for a meaningful application in chemistry and biochemistry. NMR becomes a spectroscopic method only with additional physical interactions that lead to a shift or a splitting of the energy levels of the nuclear spins and the corresponding transition frequencies. Hence, for a single isotope several resonance absorption lines instead of just one may be obtained, whose positions in the spectrum are related to molecular properties. The most important of these physical interactions will be discussed in the following paragraphs.

1.2.1 Chemical Shift

The resonance condition as given in (1.3) is strictly valid only for isolated nuclei in free space. However, in reality the nuclei are situated in matter as parts of molecules and are surrounded by an electronic shell. This electronic shell shields the nuclei somewhat with respect to the external magnetic field B_0. Thus the nucleus is not in a magnetic field B_0 but in an effective magnetic field B':

$$B' = B_0(1 - \sigma) \quad , \tag{1.28}$$

where σ is the shielding constant. Hence, the effective magnetic field has to be inserted into the resonance condition

$$\omega = \omega_I = \gamma_I B' = \gamma_I B_0(1 - \sigma) \quad . \tag{1.29}$$

Since this shift of the resonance frequency depends on the chemical surroundings into which the nucleus is placed, it is termed a chemical shift. It is usually not measured absolutely but instead relative to a standard substance S in parts per million (ppm), and is expressed by a dimensionless constant δ which is defined as the difference of the shielding constants σ or as the difference of the resonance frequencies $\omega = 2\pi\nu$ in an external magnetic field B_0

$$\delta = \sigma_S - \sigma = \frac{\nu - \nu_S}{\nu_S} \quad , \tag{1.30}$$

where σ_S and ν_S are the shielding constants and the resonance frequency, respectively, of the standard. Thus a positive δ means that the resonance of the nuclear spin investigated occurs at a higher frequency than that of the standard.

For ^1H nuclei that are surrounded by just one electron this shift is of the order of 10 ppm; for ^{13}C and ^{31}P it amounts to several hundred ppm. Although it is rather small, the chemical shift is the most important parameter of high resolution NMR. It permits one to observe selectively single nuclei in the molecule as well as groups of nuclei, and to draw conclusions about their chemical surroundings. Thus in the first NMR spectrum of ethanol shown above (Fig. 1.1) the proton resonance signals of the CH_3, CH_2 and OH groups can be easily distinguished by their $3 : 2 : 1$ intensity ratios.

However, the chemical shift is not the only important parameter which modifies an NMR spectrum. The NMR spectrum of a nuclear spin is also influenced by the interaction with the magnetic moments of neighboring nuclear spins.

1.2.2 Dipole-Dipole Coupling

The interaction between the magnetic moments of neighboring spins leads in principle to a splitting of the NMR absorption lines. The effect of a magnetic dipole μ_I on a neighboring dipole is equivalent to that of an additional magnetic field. The nuclear spin of a neighboring nucleus behaves as if it were placed not only in the external magnetic field B_0, but also in a local field B_{loc}. Its component parallel to the B_0-field is given by,

$$B_{loc} = \frac{\mu_0}{4\pi} \frac{\mu_z(3\cos^2\theta - 1)}{r^3} \quad . \tag{1.31}$$

Here, r is the distance between the two dipoles, μ_z the component of the magnetic moment μ_I parallel to B_0, and θ the angle between the magnetic field B_0 and the vector r that connects the two dipoles. As can be seen from (1.31), the magnetic dipole coupling decreases very strongly with the third power of the distance r between the two dipoles; furthermore, it is anisotropic and vanishes completely for $\cos^2\theta = 1/3$. The corresponding angle θ of about 55° where B_{loc} equals zero is called the "magic angle".

At other angles in single crystals and in simple cases of polycrystalline material a splitting of the NMR absorption lines is observed. However, in addition to the coupling between two neighboring spins, another coupling exists with other nuclei in the same molecule or in neighboring molecules that is smaller, because of the larger distance r, but not vanishing. In polycrystalline or amorphous materials a broadening of the absorption lines instead of a splitting is therefore often observed. Another important application of NMR in single crystals is based on the strong dependence of the dipole-dipole coupling on the distance between the nuclear spins concerned because of the factor $1/r^3$ (1.31): the distance between two nuclei can be measured with high accuracy, which is particularly important for ^1H nuclei because it cannot be measured accurately by means of structure determination with X-ray and electron diffraction.

In distinction to the solid state, the dipole-dipole coupling in liquids does not lead to a splitting of the absorption lines. Here, the Brownian molecular motion causes a fast reorientation of the molecules with respect to the direction of the magnetic field; the time average of B_{loc} becomes zero and the dipolar interaction is averaged out. However, in the case of large molecules of biological interest this thermal motion is not fast enough for a complete averaging out of the dipole-dipole interaction and the remaining part contributes to the line width.

1.2.3 Indirect Spin-Spin Coupling

Although the dipole-dipole interaction does not cause a splitting of the NMR absorption lines in liquids, in many cases such a splitting is observed. It results from the indirect spin-spin coupling, an interaction that is transferred between two spins by the binding electron pair. Other terms for indirect spin-spin coupling are J-coupling and, since the indirect spin-spin coupling is almost always orientation-independent, scalar coupling. The underlying mechanism of indirect spin-spin coupling is explained as follows using the example of the hydrofluoric acid molecule HF.

If one of the binding electrons of the pair e_1 is at a given moment close to proton A, the probability of a parallel orientation of the spins of e_1 and A is higher than that of an antiparallel orientation, since the electron possesses a negative moment and the proton a positive one. The same is valid for the other electron e_2 and the nucleus X of the fluorine atom. Because of the Pauli exclusion principle the spins of the two electrons must be oriented antiparallel with respect to each

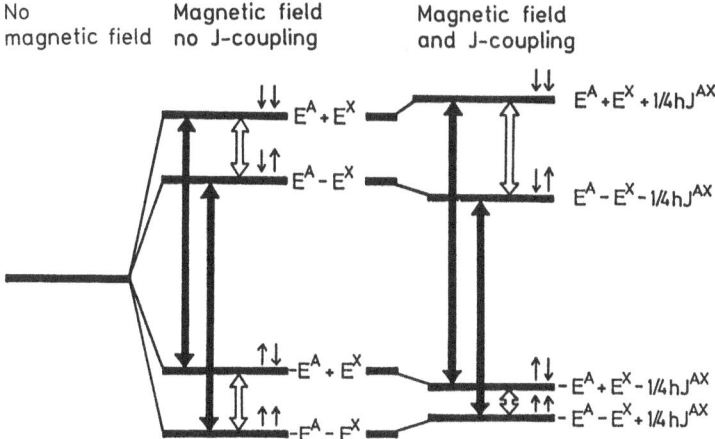

Fig. 1.9. Energy level splitting by *J*-coupling. The additional splitting of the energy levels by the *J*-coupling between nuclear spin *A* and nuclear spin *X* is not drawn to scale. It is usually several orders of magnitude smaller than the Zeeman splitting by the external magnetic field. The resonance frequencies of spin *A* and *X* are symbolized by ($\leftrightarrow$) and ($\Leftrightarrow$) respectively

other; therefore a coupling results between the nuclear spins A and X. Since in this coupled system the energy of the states depends on the orientation of the two nuclear spins, a splitting of the energy levels is observed (Fig. 1.9). The resulting differences of the transition frequencies lead to two resonance absorption lines, that is, to a splitting of the lines.

This scalar interaction is independent of the size and the direction of the external magnetic field B_0. It is measured in terms of the spin-spin coupling constant J, usually in frequency units, Hz. Its value decreases with increasing number n of transferring chemical bonds between the nuclei concerned and usually becomes immeasurably small for $n > 3$. Figure 1.10 shows schematically an example of J-coupling by the NMR spectrum of a somewhat more complicated substance, PF$_3$. The NMR spectrum of the ^{19}F nuclear spins is split into two absorption lines of equal intensity by the coupling of the ^{31}P nucleus with spin $I = 1/2$. The equidistant quartet of the ^{31}P-NMR absorption lines originates from the coupling of the resulting total spin $I_{max} = 3/2$ of the three equivalent ^{19}F spins each with spin $I = 1/2$. The distance between the four components is determined by the same spin-spin coupling constant $J = 1.44$ kHz as in the case of the ^{19}F resonance. The intensity ratio of 1:3:3:1 follows the binomial coefficients. HF and PF$_3$ are examples of weakly coupled systems in which the coupling J is much smaller than the differences of the resonance frequencies. In this case the splitting and the intensities of the lines follow relatively simple rules. The coupling between similar nuclear spins in chemically equivalent positions as represented by the three fluorine spins in PF$_3$ in principle never leads to a splitting of the absorption lines. The splitting pattern can easily be obtained:

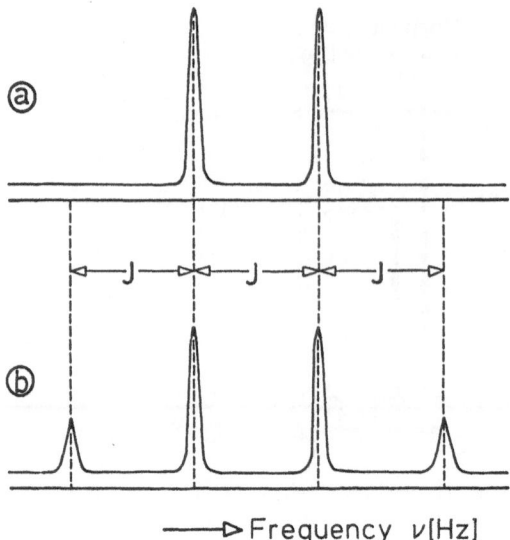

Fig. 1.10a,b. NMR spectrum of PF₃. The corresponding ¹⁹F-NMR (a) and ³¹P-NMR spectra (b) are plotted schematically. The absolute resonance frequencies are, of course, very different; only the splittings of the lines correspond to each other

Table 1.2. Number ($2NI + 1$) and intensities of the resonance lines in the case of coupling with N equivalent nuclei with spin I

$2NI+1$		intensities
2		1 1
3		1 2 1
4	$I = 1/2$	1 3 3 1
5		1 4 6 4 1
6		1 5 10 10 5 1
3		1 1 1
5		1 2 3 2 1
7	$I = 1$	1 3 6 7 6 3 1
9		1 4 10 16 19 16 10 4 1
4		1 1 1 1
7		1 2 3 4 3 2 1
10	$I = 3/2$	1 3 6 10 12 12 10 6 3 1
13		1 4 10 20 31 40 44 40 31 20 10 4 1
6	$I = 5/2$	1 1 1 1 1 1
11		1 2 3 4 5 6 5 4 3 2 1
16		1 3 6 10 21 25 27 27 25 21 15 10 6 3 1
21		1 4 10 20 35 56 80 104 125 140 146 140 125 104 80 56 35 20 10 4 1

the coupling of spin A with a spin 1/2 particle X leads to two lines of half intensity at a distance J_{AX}. If there are more coupled spins, for instance a spin M with a coupling constant J_{AM}, each of these two lines is again split into a doublet with a splitting J_{AM}. This is particularly simple in the case of the coupling with equivalent nuclear spins, as with PF_3, because here the coupling constants J are all identical. For an AX_N system, that is for the coupling of A with N equivalent nuclear spins X, $N+1$ resonance absorption lines with a distance J_{AX} are obtained; their intensities correspond to the binomial coefficients. Couplings with a nucleus with a nuclear spin I that is larger than $1/2$ lead, not to a doublet splitting, but to a multiplet of $2I+1$ absorption lines, for instance for deuterium ($I=1$) to a triplet, where the intensity ratios have to be calculated for each particular case (Table 1.2).

If the difference in the resonance frequencies is comparable with the J coupling (the strong coupling case), the coupling pattern does not follow these simple rules. We shall refer to this again later.

In the case of PF_3 the coupling between phosphorus and fluorine is comparatively large, in the range of MHz. For ^1H nuclear spins the coupling constant J is much smaller; for the ^{13}C-^1H coupling in an aliphatic ^{13}C-^1H fragment $J \approx 125\,Hz$ is found. In ethanol, where the spin-spin coupling between the protons of the $-CH_2$ and the $-CH_3$ groups extends over three bonds the coupling constant J is only $7\,Hz$.

1.2.4 Nucleus-Electron Interaction

The magnetic resonance of atomic nuclei can be influenced by the coupling with the magnetic moments μ_S of unpaired electrons with spin $1/2$ in addition to the coupling with neighboring nuclear spins. In distinction to the nuclear spin I the electronic spin is usually denoted by S. Unpaired electrons are present in free radicals and many ions that sometimes form part of macromolecules of biological interest and sometimes are attached to them by chemical methods (spin labels). This interaction is very strong because the magnetic moment μ_S of the electrons is larger by a factor of about 1000 (for protons $\mu_S/\mu_I = 657$, for other nuclei with a smaller magnetic moment μ_I the ratio is correspondingly larger). The interaction is based on two mechanisms. Firstly, there is the classical dipole-dipole coupling mentioned above that obeys (1.31), with the difference that in the numerator there is the much larger magnetic moment μ_S of the electron instead of μ_I. Furthermore, there exists a second type of interaction, namely the Fermi contact interaction. This is due to the electron in a molecule or ion having a spatial distribution of its charge and spin density that is described by its molecular orbital. If this has a final value at the position of the investigated nucleus, a coupling between the two can be observed which is isotropic and is described by a scalar coupling constant a measured in Hz.

Observing the magnetic resonance of unpaired electrons by electron spin resonance (ESR), one finds that the coupling with the magnetic moments of the atomic nuclei in general leads to a very complex splitting known as hyperfine structure, which will not be further discussed here. This type of splitting of the absorption lines does not occur in NMR. The electron spins flip so fast to and fro between the orientations with the magnetic quantum numbers $m_S = +1/2$ and

−1/2, because of their much faster spin-lattice relaxation, that the nuclear spin "sees" only an average value. This value is, however, not equal to zero because always somewhat more electron magnetic moments μ_S are oriented parallel to the direction of the magnetic field than antiparallel, analogous to the explanation given above for the nuclear moments μ_I. The resulting macroscopic magnetic moment of the electron spins causes the paramagnetism of free radicals and ions; the interaction with the nuclear spins leads to a paramagnetic shift and contributes to the relaxation. This will be explained in Sect. 1.3.7.

1.2.5 Nuclear Quadrupole Interaction

All nuclei with spin $I \geq 1$ possess, in addition to their magnetic moment μ_I, an electrical nuclear quadrupole moment eQ, which measures the deviation of the distribution of the nucleus' positive charge from spherical symmetry. Here, e is the elementary charge and $eQ = \int r^2 (3\cos^2\theta - 1)\varrho \, d\tau$ has a positive sign for a cigar-shaped nucleus and a negative sign for a flattened disk-shaped one (ϱ is the charge density, r the distance of the volume element $d\tau$ from the origin and θ the angle between the corresponding radius vector r and the axis of quantization of the spins).

In non-cubic crystals and in almost all molecules the nuclei are situated in an inhomogeneous electric field. The corresponding field gradient is defined as the second derivative of the potential V in the direction of the spatial coordinates. The largest component of the field gradient, which is in most cases oriented parallel to the chemical bond, is given by $eq = \partial^2 V/\partial z^2$. In such an inhomogeneous electric field a nucleus with a quadrupole moment can only have a well-defined small number $(2I + 1)$ of orientations each of which correspond to a discrete energy level. The energy of these levels is proportional to the product $e^2 Qq$ of the quadrupolar moment eQ and the field gradient eq. Hence, this product is a directly measurable quantity.

In the case of pure quadrupole resonance without an external magnetic field the energy levels E_Q are determined by an electrical interaction. Magnetic dipole transitions analogous to NMR can be induced by a magnetic radio frequency field that satisfies the resonance condition $\hbar\omega = \Delta E_Q$. Pure quadrupole resonance is not relevant for investigating biomolecules. Hence, we are here more interested in the additional influence of the nuclear quadrupole interaction $e^2 Qq$ on the NMR spectrum. This is explained in Fig. 1.11 by the example of the ^{14}N nucleus with nuclear spin $I = 1$. This nuclear spin can orient itself with respect to an external magnetic field B_0 in three different orientations corresponding to the magnetic quantum numbers $m_I = +1, 0, -1$. Without quadrupole interaction the splitting between these three energy levels would be equidistant and only one NMR absorption line would be observed. However, owing to the nuclear quadrupole interaction, the two energy levels with $m_I = \pm 1$ (independent of the sign) are shifted in one direction and the energy level with $m_I = 0$ in the opposite direction, so that the splitting is no longer equidistant and two absorption lines are observed. In solids the quadrupole constant $e^2 Qq$ can be measured from the

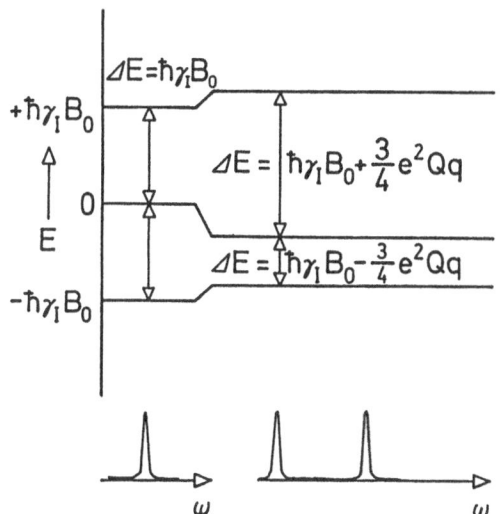

$\Delta E = \hbar \gamma_I B_0$

$+\hbar \gamma_I B_0$

0

E

$-\hbar \gamma_I B_0$

$\Delta E = \left| \hbar \gamma_I B_0 + \dfrac{3}{4} e^2 Qq \right|$

$\Delta E = \left| \hbar \gamma_I B_0 - \dfrac{3}{4} e^2 Qq \right|$

ω ω

Fig. 1.11. Nuclear quadrupole interaction. Splitting of the energy levels by nuclear quadrupole interaction (*above*) and corresponding spectra (*below*). The energy splitting varies with the orientation of the quadrupole moment with respect to the magnetic field B_0. Shown here is the orientation when the quadrupole splitting is at its maximum

separation of these two lines as well as from the pure quadrupole resonance; in liquids it just causes an undesired broadening of the NMR absorption lines.

1.3 Relaxation Mechanisms

In the preceding chapters we considered the interactions that define the position of the resonance absorption lines in the NMR spectrum. However, a real NMR spectrum is not composed of infinitely narrow absorption lines, but of lines with a finite line width. The width of an individual, homogeneous line is determined by the transverse relaxation time T_2. The longitudinal relaxation time T_1 also influences the NMR spectrum; it essentially affects the observed line intensities.

Transverse and longitudinal relaxations are induced by processes at the molecular level. They represent the interactions of the nuclear spins with their surroundings. The relaxation rates are proportional to the square of the strength of these interactions. In the case of spin-lattice relaxation, where energy must be exchanged with the surroundings, these interactions must be time-dependent in such a manner that they contain the frequency component at the Larmor frequency ω_I. The same mechanisms also influence the spin-spin relaxation, since the T_1 processes destroy the phase coherence. However, the time modulation of the interactions is not a necessary precondition for the destruction of the phase coherence, and time-independent processes also contribute to the transverse relaxation.

Five interactions are important for the nuclear magnetic relaxation in biological systems:

1. The dipole-dipole interaction.
2. The chemical shift anisotropy.

3. The indirect spin-spin interaction.
4. The nuclear quadrupole interaction.
5. The interaction with unpaired electrons (free radicals and paramagnetic ions).

Another mechanism of spin-lattice relaxation, the spin-rotation interaction, can contribute significantly to the relaxation in small molecules in gases and in low viscosity liquids. For almost all molecules of biological interest it is negligible.

1.3.1 Transverse Relaxation and Line Width

The transverse relaxation time T_2 measures the line width directly. The shape of a homogeneous magnetic resonance absorption line is described by the Lorentzian function $L(\omega)$ (Fig. 1.12),

$$L(\omega) = \frac{\Delta\omega_{1/2}}{(\Delta\omega_{1/2})^2 + (\Delta\omega)^2} \quad , \tag{1.32}$$

where $\Delta\omega = \omega_I - \omega$ and $\Delta\omega_{1/2}$ equals half the line width at half height. The relation between the line width $\Delta\omega_{1/2}$ and T_2 is given by

$$\Delta\omega_{1/2} = \frac{1}{T_2} \quad . \tag{1.33}$$

Choosing the frequency ν instead of the angular frequency ω leads to

$$\Delta\nu_{1/2} = \frac{1}{2\pi T_2} \quad . \tag{1.34}$$

However, Lorentzian-shaped lines are observed experimentally only approximately, since interactions within the sample and a remaining inhomogeneity of the magnetic field B_0, which are always present, affect the shape of the curves. Such an inhomogeneous line can in many cases be described by a Gaussian function. In cases where the lineshape is not a pure Lorentzian function, an ef-

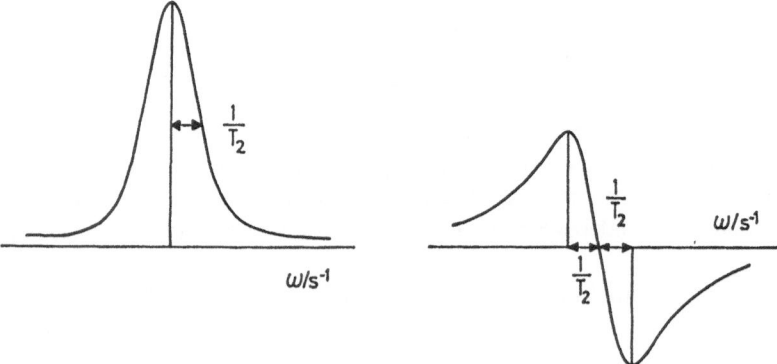

Fig. 1.12. The Lorentzian line. The typical absorption signal in a homogeneous magnetic field, the Lorentzian line (*left*), and the corresponding dispersion signal (*right*)

fective relaxation time T_2^* is, in analogy to the relation between T_2 and $\Delta\omega_{1/2}$, frequently defined by

$$T_2^* = \frac{1}{\Delta\omega_{1/2}} \ .$$
(1.35)

1.3.2 Time-Dependence of Interactions

The time-dependence of interactions in liquids is mainly due to the Brownian molecular motion. The intensity of the noise spectrum at the frequency ω_I produced in this manner determines the effectiveness of the relaxation mechanism for the spin-lattice relaxation rate $1/T_1$. The rotational correlation time τ_{rot} measures the average time in which the whole molecule, or part of it, that contains the nuclear spin is rotated by an angle of 1 rad. In low-viscosity liquids τ_{rot} is of the order of 10^{-11}s for small molecules. Consequently, the relation

$$\tau_{rot}\,\omega_I \ll 1$$
(1.36)

is satisfied. In this case the noise spectrum is "white", that is, it contains all frequencies from zero to ω_I and above with almost equal intensity. Condition (1.36) is in general not satisfied with biological macromolecules.

T_2 is always shorter than T_1, with the exception of some rare special cases. The reason is that all T_1 processes, that is, the flipping of the nuclear spin with simultaneous transfer of a corresponding energy quantum to the lattice or from the lattice, destroy the phase relations with the neighboring nuclear spins and consequently lead to transverse relaxation as well. The less relation (1.36) is satisfied, the more T_1 and T_2 differ and the greater is the inequality $T_1 > T_2$. However, the expressions given in the following paragraphs are restricted to the case of extreme narrowing where the relation (1.36) holds.

1.3.3 Dipolar Relaxation

The most important interaction for protons is the dipole-dipole coupling with similar nuclear spins, since protons possess a large magnetic moment μ_I and are in general present in high concentration, that is, with small distances between neighboring protons. Also, for other nuclei with spin $I = 1/2$, it is mostly the dipole-dipole interaction with protons that plays an important role. Quantitatively, the relaxation rate $1/T_{1,2}^{DD}$ by dipole-dipole interaction between a nuclear spin A with $I_A = \frac{1}{2}$ and a nuclear spin X with $I_X = \frac{1}{2}$ is given by

$$1/T_{1,2}^{DD} = \left(\frac{\mu_0}{4\pi}\right)^2 \frac{\gamma_A^2 \gamma_X^2 \hbar^2 \tau_{rot}}{r^6} \ .$$
(1.37)

Equation (1.37) can be easily understood: the nuclear spin A is situated in a local magnetic field produced by the nuclear spin X, which, following (1.31), is proportional to γ_X/r^3 and depends, furthermore, on the angle θ between r and B_0. The time-dependence of the dipole-dipole interaction between the two nuclear spins in the same molecule is due to the constant variation of the angle θ with respect to the external magnetic field because of the Brownian molecular motion. The rapidity of this variation is measured in terms of the rotational correlation time τ_{rot} mentioned above. If the two nuclear spins concerned belong to two different molecules, the distance r also varies by diffusion; it is measured in terms of the translational correlation time τ_{trans}, which is expressed by a small modification of (1.37). For the special case of a dipolar interaction between two magnetically equivalent nuclei (identical nuclei in identical surroundings; $\gamma_A = \gamma_X, \omega_A = \omega_X$) the right side of (1.37) has to be multiplied by the factor $3/2$.

1.3.4 Relaxation by Chemical Shift Anisotropy

A second interaction important for relaxation is the anisotropy of the chemical shift. The electron cloud produces an additional local magnetic field at the position of the nuclear spin that is almost always anisotropic. Its value therefore varies because of the Brownian molecular motion, with the time-dependence again measured in terms of the correlation time τ_{rot}. Since this interaction increases proportionally to B_0, and since the relaxation rate is, as always, proportional to the square of the interaction, a relaxation rate $1/T^{cs}$ in low-viscosity liquids is obtained within the range of validity of (1.36) for the simple case of a cylindrically symmetric chemical shift

$$1/T_1^{cs} = \frac{2}{15}\gamma^2 B_0^2 \Delta\sigma^2 \tau_{\text{rot}} \quad . \tag{1.38}$$

In this equation $\Delta\sigma$ denotes the anisotropy of chemical shift. Even for the fast rotational diffusion of small molecules ($\omega_I \tau_{\text{rot}} \ll 1$), T_1 is not equal to T_2 as it is in the case of dipolar relaxation, but

$$T_2^{cs} = \tfrac{6}{7} T_1^{cs} \quad . \tag{1.39}$$

For protons the anisotropy of chemical shift is of minor importance compared to the dipole-dipole coupling, since the anisotropy is comparatively small because it is shielded by one electron only. For other nuclei with $I = 1/2$, such as ^{13}C, ^{15}N, ^{19}F, ^{31}P, its contribution to relaxation with magnetic fields of medium strength is frequently not negligible. In the case of high magnetic fields of 10 T and more, which are not unusual in modern spectroscopy, it can become the predominant relaxation mechanism.

1.3.5 Relaxation by Indirect Spin-Spin Coupling

The scalar spin-spin coupling between two nuclear spins A and X described above can also contribute to the relaxation. In this case the time-dependence cannot be due to the rotational diffusion of the molecule, since the spin-spin coupling constant J_{AX} is independent of this motion. However, the time-dependence may originate from two different mechanisms. Firstly, the spin-spin coupling constant

J_{AX} may be time-dependent as a result of chemical exchange of one of the two nuclei, mostly a hydrogen atom. This is usually known as scalar relaxation of the first type, and the corresponding correlation time equals the reciprocal exchange rate. Secondly, the local magnetic field produced by the nuclear spin X at the position of the nuclear spin A can be modulated by the relaxation of the nuclear spin X. This case is known as scalar relaxation of the second type; the corresponding correlation time is the relaxation time of the nuclear spin X.

Scalar relaxation of the first type can produce a significant contribution to the relaxation, in particular for protons and deuterons, since these nuclear spins undergo a fast exchange more often than others. The scalar relaxation of the second type is particularly important for nuclear spins A that are coupled to a nuclear spin X with $I \geq 1$, which has a high relaxation rate because of the quadrupole interaction.

Quantitatively, the relaxation rate $1/T_1^{sc}$ of the nuclear spin A by scalar relaxation is given by

$$\frac{1}{T_1^{sc}} = \frac{8}{3}\pi^2 J_{AX}^2 I_X(I_X + 1)\frac{\tau_{sc}}{1 + (\omega_A - \omega_X)^2\tau_{sc}^2} \quad . \tag{1.40}$$

Here, I_X is the quantum number of the nuclear spin X, and τ_{sc} is either the relaxation time of the nuclear spin X or the reciprocal exchange rate. For the tranverse relaxation rate a similar relation is valid

$$\frac{1}{T_2^{sc}} = \frac{4}{3}\pi^2 J_{AX}^2 I_X(I_X + 1)\left(\tau_{sc} + \frac{\tau_{sc}}{1 + (\omega_A - \omega_X)^2\tau_{sc}^2}\right) \quad . \tag{1.41}$$

1.3.6 Quadrupolar Relaxation

For nuclei with spin $I \geq 1$ the nuclear quadrupole interaction is usually the predominant relaxation mechanism. The time-dependence is due to the Brownian molecular motion as in the case of the dipole-dipole interaction. For the most frequent case of a field gradient of approximately cylindrical symmetry along a molecular axis z the relaxation rates $1/T_{1,2}^Q$ within the range of validity of (1.36) are given by

$$\frac{1}{T_{1,2}^Q} = \frac{3}{40}\frac{2I + 3}{I^2(2I - 1)\hbar^2}e^4Q^2q^2\tau_{\text{rot}} \quad , \tag{1.42}$$

where τ_{rot} is again the rotational correlation time. Nuclei with a quadrupole moment are less suitable for high resolution NMR because of their larger line width, which is due to the fast relaxation caused by the quadrupole coupling. An exception is deuterium whose quadrupole moment is so small that a comparatively good resolution of the NMR spectrum can be achieved.

1.3.7 Paramagnetic Relaxation

If a nuclear spin $I = 1/2$ is coupled with unpaired electrons, either by dipole coupling or by isotropic Fermi coupling, this interaction usually predominates for the nuclear spin relaxation because of the much larger magnetic moment μ_S of the electrons. The time-dependence originates in this case mainly from the much shorter spin-lattice relaxation time T_1 of the electronic spin S, that is, the relevant correlation time τ_S equals the electronic relaxation time T^S. If we denote by $1/T_{1,2}^{HF}$ the relaxation rate of the nuclear spin A owing to its hyperfine interaction with the electronic spin S, it is in the simplest case

$$1/T_{1,2}^{HF} = a_{1,2}^2 S(S+1)\tau_S \quad , \tag{1.43}$$

where $a_{1,2}$ includes the scalar as well as the dipolar part of the hyperfine interaction. However, for a precise quantitative description this simple expression is inadequate.

Only rarely are such unpaired electrons present in biomolecules. Frequently, however, stable free radicals or rare earth ions are bound at well-defined positions in macromolecules for the purpose of structure investigation (spin labels), or a diamagnetic ion, like magnesium, is substituted by a paramagnetic one, like manganese. Applications of this technique are dealt with in Chap. 2. The change in the relaxation times on adding paramagnetics is the basis for the use of these substances as contrast agents in NMR tomography (Chap. 4).

1.4 Experimental Methods

In this section the experimental methods and the principles for the construction of an NMR spectrometer will be discussed. The historically older continuous-wave detection of NMR has today been almost completely replaced by Fourier spectroscopy. However, continuous-wave (cw) detection can perhaps be somewhat more easily understood, since in this case, in analogy to classical spectroscopy, the absorption of the electromagnetic quanta are observed directly as a function of their frequency (energy). In pulsed NMR spectroscopy, on the other hand, the time-dependent NMR signal, the free induction decay (FID), is detected. This time-dependent signal is converted by a mathematical operation, Fourier transformation, into a frequency spectrum that is equivalent to the spectrum with cw detection, at least in all simple cases. However, compared to cw methods, Fourier spectroscopy is superior in its sensitivity of detection and its flexibility. As we shall see, all NMR spectrometers have much in common, although a whole-body tomograph is, of course, distinctly different from a high resolution spectrometer, and not only in size.

1.4.1 Continuous-Wave Detection of NMR

With the simplest method of detection of a high resolution NMR spectrum, which was almost exclusively used during the first two decades of NMR spectroscopy, the sample is placed into a homogeneous magnetic field and a monochromatic radio frequency field is swept continuously over the range of the NMR absorption lines.

In general, the frequency ω of the B_1-field is scanned slowly through resonance and the signal is continuously detected. Usually, the absorption signal is detected, that is, the $M_{y'}$ component of the magnetization that is phase shifted with respect to the B_1-field by 90°. Alternatively, the $M_{x'}$ component in phase with B_1 can also be measured and a dispersion signal obtained. The behavior of the magnetization with this slow passage is described as a solution of the Bloch equations in the rotating coordinate system if the derivatives are set equal to zero:

$$M_{x'} = M_0 \frac{\gamma_I B_1 (\omega_I - \omega) T_2^2}{1 + (\omega_I - \omega)^2 T_2^2 + \gamma_I^2 B_1^2 T_1 T_2} \tag{1.44}$$

$$M_{y'} = M_0 \frac{\gamma_I B_1 T_2}{1 + (\omega_I - \omega)^2 T_2^2 + \gamma_I^2 B_1^2 T_1 T_2} \tag{1.45}$$

$$M_z = M_0 \frac{1 + (\omega_I - \omega)^2 T_2^2}{1 + (\omega_I - \omega)^2 T_2^2 + \gamma_I^2 B_1^2 T_1 T_2} \tag{1.46}$$

For small radio frequency powers ($\gamma_I^2 B_1^2 T_1 T_2 \ll 1$) the absorption signal takes the shape of a Lorentzian curve (1.32). The absorption and dispersion signals are plotted in Fig. 1.12.

If the rate of the energy absorption from the B_1-field is comparable to or larger than the spin-lattice relaxation rate $1/T_1$, the amplitude of the absorption signal decreases because the population difference $N^+ - N^-$ of the two levels concerned is reduced compared to its value in Boltzmann equilibrium, and simultaneously the linewidth increases, an effect which is termed "saturation". The degree of saturation depends, of course, on the one hand on the relaxation times T_1 and T_2 and on the other hand on the strength of the B_1-field. Quantitatively, it is measured in resonance by the saturation factor s.

$$s = 1 + \gamma_I^2 B_1^2 T_1 T_2 \quad . \tag{1.47}$$

The amplitude of the observed absorption signal decreases again owing to saturation at higher B_1-fields. The strongest signal is obtained with the B_1-field B_1^{opt} given by

$$B_1^{\text{opt}} = \frac{1}{\gamma_I \sqrt{T_1 T_2}} \quad . \tag{1.48}$$

However, B_1^{opt} is no longer in the linear range. The line broadening that occurs with increasing B_1-field amounts at B_1^{opt} already to a factor of $\sqrt{3}$. The field-

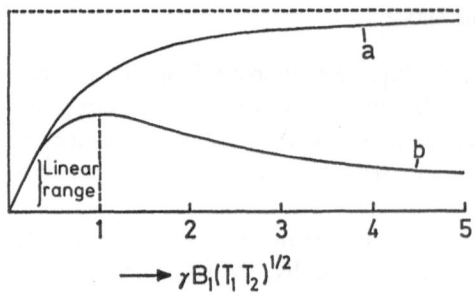

Fig. 1.13. Saturation of the NMR signal in cw spectroscopy. The total intensity (a) and the amplitude (b) of an NMR absorption line are plotted as a function of the B_1-field

dependence of the amplitude and the total intensity (area under resonance line) of the absorption signal are depicted in Fig. 1.13.

1.4.2 Pulsed NMR

The simplest way of performing an NMR pulse measurement consists of a 90° pulse which flips the magnetization **M** into the xy-plane followed by observation of the free induction decay. After the 90° pulse the FID is observed on the oscilloscope as a function $f(t)$ of the time t, that is, in the time domain. With a simple resonance signal, as in the case of the ^{1}H atoms in water, if ω equals the resonance frequency ω_I, then $f(t)$ is a simple exponentially decaying function. To this simple time domain exponential decay there corresponds a function $F(\omega)$ in the frequency domain that has the shape of a Lorentzian line. In mathematics, the two functions in the time domain and in the frequency domain are related to each other by

$$F(\omega) = \int_{-\infty}^{\infty} f(t)e^{-i\omega t} dt \tag{1.49}$$

$$f(t) = \frac{1}{2\pi} \int_{-\infty}^{\infty} F(\omega)e^{+i\omega t} d\omega \tag{1.50}$$

which are known as Fourier transforms of each other after the French mathematician Jean Baptiste de Fourier.

The typical duration of a 90° pulse in a high resolution spectrometer is of the order of 10^{-5} seconds. In contrast to a cw-irradiated radio frequency such a pulse is not monochromatic, but contains a whole frequency range on both sides of the frequency ω. The corresponding frequency spectrum is obtained by a Fourier transformation of the pulse, which is described for a rectangular pulse of width $2\Delta\tau$ by the function

$$\operatorname{sinc} \omega\Delta\tau = \frac{\sin \omega\Delta\tau}{\omega\Delta\tau} . \tag{1.51}$$

This frequency distribution becomes the broader the narrower $2\Delta\tau$ is, that is, in the limit of an infinitely narrow pulse (a δ-pulse) a white frequency spectrum is obtained that contains all frequencies.

28

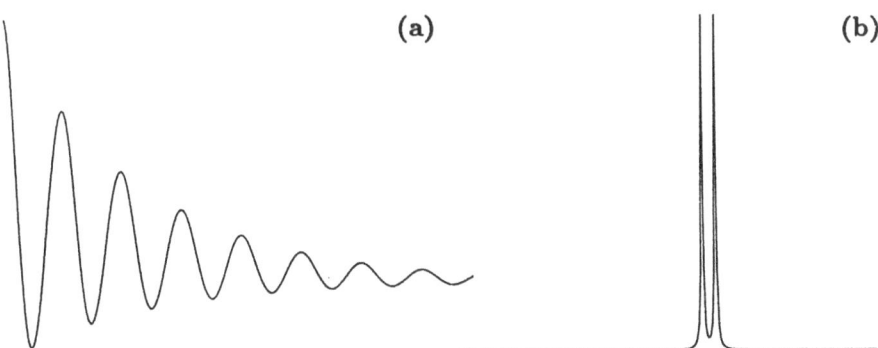

Fig. 1.14a,b. NMR spectrum in the time and frequency domain. The NMR signal of two absorption lines in the time domain (a) and in the frequency domain (b), which can be transformed into each other by Fourier transformation

If a sample contains different magnetically non-equivalent protons, all of them can be excited by such a pulse. In this case the FID is not a simple exponential function, but a superposition of many frequency-modulated exponential functions. In principle the function $f(t)$ in the time domain contains the same information as the function $F(\omega)$ in the frequency domain. However, the human eyes and brain are much more capable of "understanding" a spectrum $F(\omega)$ in the frequency domain. Figure 1.14 shows as an example a comparatively simple two-line spectrum in the time domain and in the frequency domain.

The main reason that pulsed Fourier transformation spectroscopy (FT spectroscopy) is nowadays almost exclusively used is because of its considerably higher sensitivity. This is because with cw spectroscopy only a single frequency is irradiated at a given time and consequently only those nuclear spins are excited that are in resonance at precisely this frequency. Signals are only detected when the scan is actually passing through a resonant line, a rather inefficient method from the point of view of use of time. In distinction to this method, with FT NMR all frequencies of the total spectrum are excited and detected simultaneously.

The signal-to-noise ratio can be improved by adding up a larger number of spectra. When measuring a physical quantity the signal is directly proportional to the number n of accumulated measurements, while the statistical noise increases in proportion to $n^{1/2}$ so that the signal-to-noise (S/N) ratio increases in proportion to $n^{1/2}$:

$$S/N \sim n^{1/2} \quad . \tag{1.52}$$

Since each scan that consists of an rf pulse and an FID requires about one second, it is possible to record in 10000 seconds = 2.5 hours 10000 passages; in this way one obtains after a Fourier transformation 100 times the signal-to-noise ratio compared to a single scan. On the other hand, the gain in the signal-to-noise ratio compared to a cw experiment, which requires for a many-line spectrum a comparable recording time T_A, is not quite as large as one would expect because in the slow passage one can use a much narrower rf band width, while in the case of Fourier spectroscopy the band width is determined by the extension of the total spectrum in frequency units. However, the gain in sensitivity is still considerable; quantitatively, it depends on the ratio of the frequency range of the total spectrum to the linewidth $\Delta\omega_{1/2}$ of the single lines.

To estimate the optimum signal-to-noise ratio we still have to consider the phenomenon of saturation. As explained above, the signal amplitude with a cw NMR experiment at low rf irradiation is directly proportional to the B_1-field (and consequently to the square root of the rf energy). In the case of a stronger B_1-field a deviation of the linear increase occurs because of the decrease of the population difference, which is accompanied by an additional line broadening until at high rf power no more signal can be detected.

An analogous effect must be taken into account with FT NMR. The delay between two 90° pulse cycles must be about three to four times the spin lattice relaxation time T_1 to ensure that the population difference between the two Zeeman levels concerned has reached again approximately the initial Boltzmann equilibrium. This means, when investigating solutions of biological molecules, a time typically of the order of three to five seconds. On the other hand, with Fourier spectroscopy there is no effect that corresponds to saturation broadening of cw NMR. This is another significant advantage.

To determine the number of nuclei that belong to a given line from the relative intensities of the individual NMR absorption lines there is no other choice left but to keep this distance of several T_1 between two pulse cycles. However, if just one well resolved spectrum is desired with a good signal-to-noise ratio in the shortest possible time, a considerable amount of time can be gained by reducing the pulse length. This method is simply based on the properties of the sine and cosine functions (Fig. 1.15).

Before starting a measurement, $M_z = M_0$ and $M_{y'} = 0$; after a 90° pulse with B_1 parallel to x', $M_z = 0$ and $M_{y'} = M_0$. If a considerably shorter pulse is applied, for instance a 30° pulse, owing to the properties of the trigonometrical functions $M_{y'} = 0.5 \, M_0$, that is, half as large as after a 90° pulse, while M_z merely undergoes a minor decrease from M_0 to 0.87 M_0; correspondingly, after a 15° pulse $M_{y'} = 0.26 \, M_0$ and $M_z = 0.97 \, M_0$. With a given longitudinal relaxation time T_1 and a selected repetition time T_R between two pulse cycles, there exists an optimum pulse angle at which the signal amplitude is at its

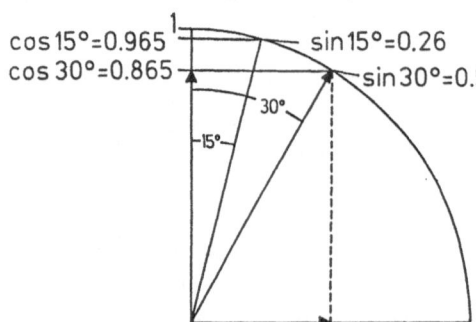

cos 15°=0.965
cos 30°=0.865
sin 15°=0.26
sin 30°=0.5

Fig. 1.15. Values of the sine and cosine functions at small angles. The values of the longitudinal and transverse components of the magnetization after a pulse with the pulse angle α are proportional to cos α and sin α. With small pulse angles one already obtains a significant transverse component when the z component is still close to the equilibrium value M_0

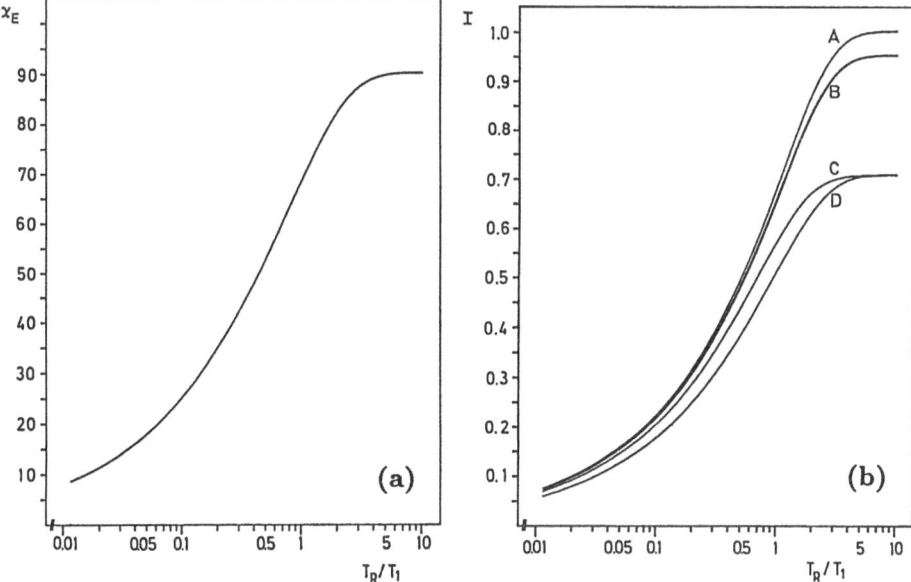

Fig. 1.16a,b. Optimum pulse angle. (a) Ernst angle α_E is plotted logarithmically as a function of the ratio of the repetition time T_R and the longitudinal relaxation time T_1. (b) Signal amplitude in dynamic equilibrium as a function of T_R/T_1 at different pulse angles. A: $\alpha = \alpha_E$, B: $\alpha = \alpha_E \pm 20\%$, C: $\alpha = \alpha_E + 50\%$ D: $\alpha = \alpha_E - 50\%$

maximum in dynamic equilibrium after several pulses. This pulse angle, which gives the optimum signal-to-noise ratio, is usually called the Ernst angle α_E (Ernst and Anderson, 1966) and is given by

$$\alpha_E = \arccos e^{-T_R/T_1} \quad . \tag{1.53}$$

In Fig. 1.16 this dependence of the optimum pulse angle on the repetition time T_R and on the longitudinal relaxation time T_1 is shown graphically. However, in a real sample different resonance absorptions may have different T_1 values, while the signal-to-noise ratio at a given T_R can be optimized for a single T_1 value only. In reality this is usually not so important, since the loss in sensitivity caused by varying the ratio T_R/T_1 by a factor of two is rather small.

Before the cw NMR signal can be processed by the computer it must first be digitalized, that is, at fixed time increments Δt the induction voltage is measured and converted into digital values. The size of the time increment Δt is given in the Nyquist theory by the highest frequency ν_{max} which is to be observed

$$\Delta t = \frac{1}{2\Delta\nu_{max}} \quad . \tag{1.54}$$

Equation (1.54) can be understood as follows: in order to distinguish the frequency of an oscillation from lower frequencies, at least two data points are required within one period of oscillation.

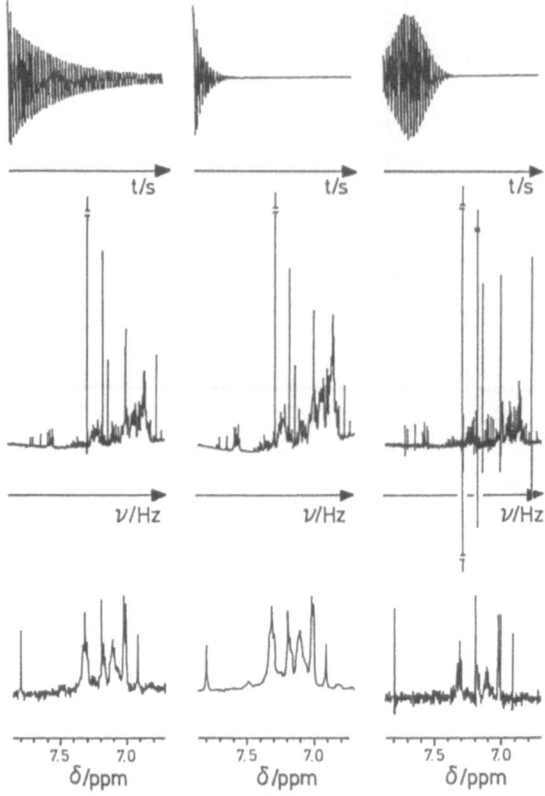

Fig. 1.17. Digital filtering of the NMR signals. (**Upper row**) ^{1}H-NMR signal (500 MHz) of HPr protein unfiltered (*left*), exponentially filtered by multiplication of the FIDs with the function e^{-at} (*middle*), Gaussian filtered by multiplication with the function e^{at-bt^2} (*right*). The FID consists of 16 384 (16k) points that were taken at intervals of 96 μs. (**Middle row**) Fourier transform of the time signals shown. Total frequency range 5208 Hz (10.4 ppm). (**Lower row**) Small section of the total spectrum. The ppm scale refers to the internal standard 2,2-dimethyl-2-silapentane-5-sulfonic acid (DSS)

Frequently, the data measured are filtered by multiplication with a suitable function before Fourier transformation. This is done for two different purposes: to reduce the noise that is always present or to increase the resolution by changing the lineshape of the resonance absorption. A simple and frequently applied filter function is multiplication by a decreasing exponential function, which leads to a small line broadening together with the suppression of the noise (Fig. 1.17). The lineshape of the Lorentz line is here preserved because the multiplication of two exponential functions results again in an exponential function. To reduce the line width a Lorentzian-to-Gaussian transformation is often used that transforms the Lorentzian absorption lines into narrower lines that are similar to Gaussian absorption lines.

The digitally stored FID is not a complete description of the unperturbed induction decay because it contains discrete values only. Correspondingly, the

Fourier transformation occurs in a computer with a fast algorithm, the fast Fourier transformation (FFT) which is adapted to the discrete form of the data.

Although the continuous Fourier transformation of the whole FID may produce an ideal frequency spectrum of the sample, in recent years the question has been more and more frequently discussed whether or not a better procedure exists to obtain the desired frequency signal from the time signal. As a matter of fact, the FID is in reality merely observed for a limited acquisition time T_{aq}, and then the measurement is repeated. The discrete Fourier transformation processes only this information and furnishes a frequency spectrum that corresponds precisely to this truncated FID and only approximates the ideal frequency spectrum that is wanted.

Two different types of data processing are today being discussed that in principle should bring about better results, since both make a better use of the information available for predicting how the observed FID would continue after the acquisition time T_{aq}. One of them is the maximum-entropy method (MEM), which was originally developed for evaluating geological data but has recently shown remarkable success in image processing; the second type comprises various optimizing methods based on the linear prediction (LP). The LP methods derive from the fact that an ideal NMR signal can be described as a superposition of exponentially damped cosine functions. If the function that produces the observed FID is found, then the FID can, of course, be predicted for any given time. However, both methods have a serious drawback: they require a large amount of computing time and are at present only applicable in special cases.

1.4.3 The NMR Spectrometer

In principle, all NMR spectrometers have a similar set-up simply because their construction has to be adapted to the requirements of the nuclear magnetic resonance experiments. The details of a cw spectrometer are shown in Fig. 1.18 and those of a pulsed spectrometer in Fig. 1.19. Even the NMR tomograph, which cannot be regarded as a spectrometer in the true sense of the word, is built according to the scheme of the Fourier spectrometer. In the following we shall have a closer look at the various components and their functions.

An indispensable part of any NMR spectrometer is the magnet that is to produce a static magnetic field B_0 as homogeneous and stable as possible. It is the B_0-field that produces the splitting between the nuclear spin energy levels and the NMR transitions induced therein.

Static magnetic fields can be produced by three different types of magnets: electromagnets, permanent magnets and superconducting magnets. With high resolution spectroscopy, superconducting magnets are nowadays almost exclusively used because only they can produce the high magnetic fields required for optimum resolution and sensitivity. In tomography, on the other hand, in addition to superconducting magnets, electromagnets are still being used as well because here it is not so clear which magnetic field is the most suitable one; furthermore,

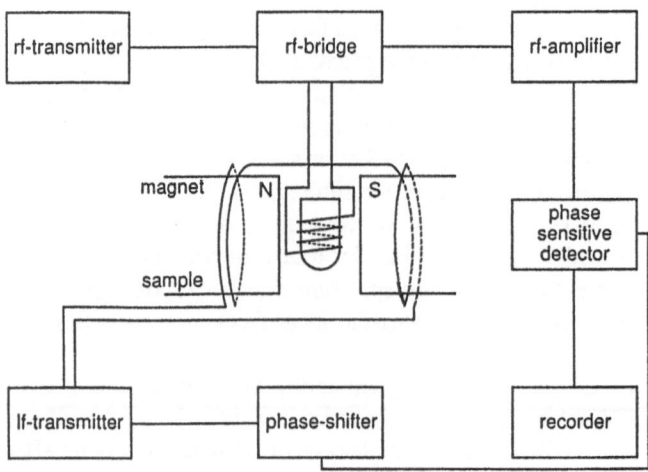

Fig. 1.18. Block diagram of a cw spectrometer

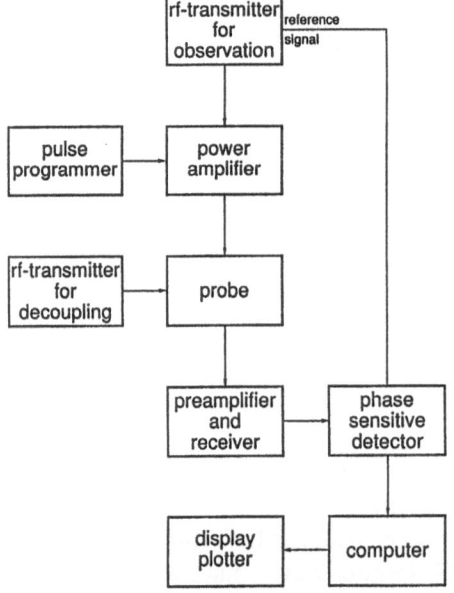

Fig. 1.19. Block diagram of a simple Fourier spectrometer

economic considerations play a major role. Permanent magnets are at present only of minor importance both for NMR spectroscopy and tomography.

The strength of the magnetic field that can be reached depends on the size of the sample investigated, and this may vary considerably. The diameter of the sample in high resolution NMR spectroscopy is of the order of 0.5–2.5 cm, while the dimension of the sample investigated by NMR tomography is given by the size of the human body. The maximum field strength of electromagnets is about 2.5 T for small volumes and about 0.3 T for large volumes. The corresponding limits for superconducting magnets reached with the present technology are about 14 T and 4 T, respectively.

Independently of the type of magnet, the requirements as to homogeneity are very high for high resolution NMR investigations of molecular structure. At the present state of the art one expects a resolution of about 0.1 Hz with a proton resonance frequency of 600 MHz, that is, 2×10^{-10}. In principle, such a homogeneity could be achieved with superconducting coils alone, but in reality this is not possible because of the variation in susceptibility of the different samples. Using additional resistive coils, which are placed between the superconducting main coil and the sample (shim coils), the field may be corrected to a homogeneity of a few 10^{-9} over the volume of the sample. A small remaining inhomogeneity in the plane perpendicular to the B_0-field can be averaged out by a mechanical rotation of the sample around the z-axis.

For wide-line cw spectroscopy, the strength of the B_0-field is also made to vary periodically, to impose a time-dependence on the NMR signal. This time-dependence is a precondition for the detection method usually used, phase-sensitive detection (Fig. 1.18). This is not necessary for Fourier spectroscopy because here the FID is already time modulated. NMR tomography requires additional gradient coils, which superpose additional field gradients on the homogeneous magnetic field.

To induce NMR transitions a radio frequency field B_1 is also needed perpendicular to the static magnetic field. The rf frequency is generated by a transmitter and applied to the sample by using a coil or a resonator. With cw spectroscopy the radio frequency is applied continuously with low power, while in the case of pulsed NMR intense pulses must be produced for short periods of the order of several microseconds. The NMR signal is then detected either by the transmitter coil or by a separate receiver coil. This extremely weak NMR signal of typically $10^{-6} - 10^{-1}$ V has to be amplified before it is phase-sensitively detected. With cw spectroscopy the signal can now be recorded directly, while with Fourier spectroscopy it has to pass through an analogue-to-digital converter (ADC) before it is fed into a computer. It is then Fourier transformed and processed, and can finally be shown on a graphic display or plotted by a recorder. An NMR pulsed spectrometer comprises a computer for controlling the spectrometer and for acquiring data. Moreover, a high performance computer is required for processing the data (fast 2D transformation, reconstruction of the 2D and 3D images). In the following we list some technical data that constitute, at the present state of the art, the minimum requirements for providing adequate working conditions, that is, for avoiding unacceptable time delays.

The CPU (central processing unit) should be a 32-bit computer with a computing speed of at least 4 MIPS (million instructions per second), a memory of 8 Mbyte or more and a disk of at least 150 Mbyte. A magnetic tape with typically 40 Mbyte is required for safeguarding the data. Optical disks with significantly higher capacity will probably gain in importance in the future. With this equipment, data processing is possible but still rather slow. One way to speed up computing by one to two orders of magnitude, depending on the type of application, is the use of an array processor. An alternative is a special fast floating point processor. However, for the important application to 2D Fourier transformation, most CPUs do not reach the speed of an array processor. A fast connection between the spectrometer and the computer is necessary because of the large size of data sets. In addition, high resolution color graphics are required, also with a fast connection to the computer.

Of course, the necessary software for these applications (the system software and frequently also the application software) is in general provided by the manufacturers. If the computer is to be used by several users, for instance for simultaneously processing several data sets (as a multiuser workstation), adequate operating systems are required, and for developing one's own programs the appropriate compilers must be available.

2. NMR Spectroscopy in Biochemistry

In the last decade nuclear magnetic resonance has, almost unnoticed, become an important method in the field of biochemical and biophysical research. NMR spectroscopy is a very powerful analytical method in biochemistry as well as in organic chemistry. It is suitable for determining the chemical structure of unknown substances and for confirming or disproving the structure of newly synthesized compounds. In addition to this chemically oriented application, NMR can also furnish information on the spatial arrangement of atoms and their motion in biologically important molecules and molecular complexes; furthermore, it can contribute substantially to the elucidation of enzyme mechanisms and of biochemical reaction pathways (Table 2.1).

An advantageous property of NMR that also plays an important role in medical applications is the low energy of the radio frequency quanta used. Even with

Table 2.1. Biochemical applications of NMR

---

1. NMR on small molecules

Resolution and assignment of all resonance lines relatively easy.

- Chemical analysis
- Determination of unknown chemical structures.
- Non-invasive determination of concentrations.
- Study of reaction pathways and reaction products.
- Determination of binding constants.
- Determination of velocity constants.
- Conformation of substrates at the active center.

2. NMR on macromolecules

Resolution and assignment of all resonance lines often very difficult

- Conformational changes by external factors (pH, temperature, ionic strength, pressure, etc.)
- Conformational changes with substrate binding
- Structure of the active center of enzymes, spatial arrangement of side chains relative to the substrate, ionization state of functional groups
- Estimation of the time scale (correlation time) and the motion type of local or global conformational changes
- Interaction between molecules, enzyme-substrate, protein- protein and protein-nucleic acid interaction
- Determination of the secondary structure in solution
- Determination of the tertiary structure in solution

---

the highest magnetic fields applied nowadays, their energy of about 4×10^{-25}J is considerably below the energies required to crack a covalent chemical bond, with typical bond energies of the order of 10^{-17}J. Hence, NMR is a non-destructive method in contrast to other methods that use ionizing radiation.

2.1 NMR as Analytical Method

In this section we shall show with examples how NMR may be applied as an analytical method in biochemistry. For this purpose some concepts that were already introduced in Chap. 1 have to be discussed here in more detail.

A typical property of NMR as an analytical tool in liquids may be demonstrated with the NMR spectrum of a simple substance, ethanol, which can be clearly recognized even in the historical NMR spectrum of this molecule shown in Chap. 1 (Fig. 1.1): three separated NMR resonance absorption lines are visible in the spectrum which correspond to the protons of the three functional groups of ethanol, namely the methyl methylene, and hydroxyl groups. This shows that the NMR spectrum is not related in some complex manner to the chemical structure, but that individual atoms or groups of atoms may be observed selectively. This distinguishes NMR essentially from many other competing analytical methods.

NMR is a comparatively insensitive method of detection even with the highest magnetic fields at present available in commercial spectrometers of about 14 T, corresponding to a proton resonance frequency of 600 MHz. The minimum amount of substance required for each NMR detection varies very strongly with the experimental conditions. With small molecules the minimum amount of substance necessary is typically about 10 nmol, an amount that can be easily detected with the simple biochemical laboratory method of thin layer chromatography. With other widely used methods, such as gas chromatography or radioimmunoassay, the femtomole or even attomole range can be reached. Hence these methods are often superior to NMR because of their higher sensitivity if it is just a matter of detecting a well-known substance. The strength of NMR first becomes apparent when NMR-typical additional information is needed.

This is always the case when the substance investigated has so far been unknown. The structural hypotheses may be directly derived from the NMR spectrum and finally the correct chemical structure can be determined. A further advantage of NMR spectroscopy, which with some restrictions is also valid for certain chromatographic methods, is its low selectivity of detection. Since for example with ^{1}H-NMR all molecules that contain protons give a signal, NMR detects also substances that are not expected to be present in the sample and that can only be found by more specific methods when they are especially looked for.

2.1.1 Identification of Known and Unknown Substances

The analytical application of NMR consists in its simplest form of the comparison of the spectrum of an unknown sample with a set of standard spectra of known substances. Agreement of the standard spectrum with the spectrum of the unknown sample leads to the identification of the substance. This is always connected with a qualitative statement on the concentration which in its simplest form is: Substance A is or is not present in the sample in a detectable concentration. This simple approach is only justified if the two spectra to be compared are taken under identical conditions. However, the actual spectrum that a substance produces depends on the experimental conditions and may strongly vary with them. The form of the spectrum is influenced not only by external parameters, such as temperature, magnetic field strength and pulse sequence that can be directly controlled by the experimentalist, but also by the precise composition of the solution that contains the substance investigated. The matter has therefore to be looked at from a more abstract point of view and characteristic properties have to be compared in order to obtain well-defined results.

Figure 2.1 shows the NMR spectrum of ethanol measured with a modern high-field spectrometer. Striking features that characterize the spectrum are the line positions, the multiplet structure and the relative intensities of the resonance absorption lines. The corresponding physical interactions have already been discussed in Chap. 1. The line position is determined by the chemical shift and the splitting of the lines by the J-coupling. The areas below the resonance lines that are obtained by numerical integration (Fig. 2.1) are proportional to the number of nuclear spins.

As mentioned above, the chemical shift may be understood as an amplification or attenuation of the external magnetic field at the nucleus by local fields. A significant component of these fields is caused by the electronic shell of the molecule concerned and is hence connected with its chemical structure. It is

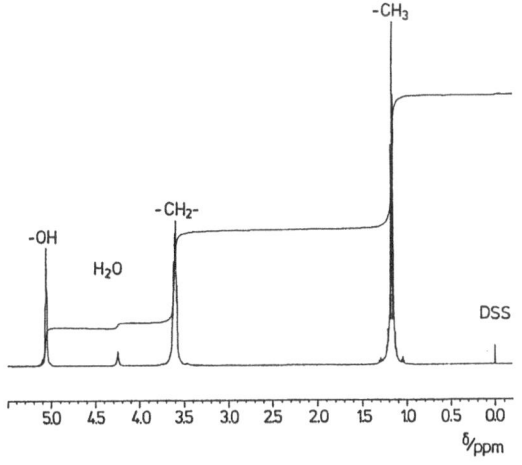

Fig. 2.1. ^{1}H-NMR spectrum of 95 % ethanol recorded at 500 MHz. The signal at 0 ppm is the CH$_3$ resonance of the standard DSS to which the ppm scale usually refers. Simultaneously depicted is the integral of the spectrum

Table 2.2. Chemical shifts in ^{1}H and ^{13}C-NMR

Group	range of chemical shifts δ relative to TMS in ppm	
	^{1}H-NMR	^{13}C-NMR
-C-CH$_3$ (ordinary methyl)	0.1 – 1.8	5–35
-S-CH$_3$ (thiomethyl)	1.5 – 3.8	10–20
-O-CH$_3$ (methoxy)	2.6 – 4.2	45–60
-C-CH$_2$-C (methylene)	0.3 – 3.9	15–55
-S-CH$_2$-C (methylene)	1.6 – 4.8	25–45
-O-CH$_2$-C (methylene)	3.2 – 4.8	40–70
-C$_2$-CH-C- (methine)	0.8 – 2.6	30–60
-C$_3$-C-C- (tertiary C)–	–	25–50
-CH = CH-C- (alkenes)	5.0 – 7.0	105–145
-CH = CH-C- (aromatics)	5.0 – 10.0	115–145
-C-CH= N- (heterocycles)	5.0 – 10.0	145–155
-C≡ C-C- (alkynes)	–	75–95
-C-CHO (aldehyde)	9.4 – 11.0	185–210
-C-COOH (acid)	4.6 – 14.0	170–185
-C-CO-NH (peptide bond)	6.5 – 11.0	170–185

reasonable to assume as a first approximation that there exists something like a hierarchy of interactions which somehow correspond to the number of bonds between the atom investigated and a substituent where the influence on the chemical shift decreases with increasing number of intervening bonds. In addition, the distribution of the electrons and hence the NMR spectrum also depends on the spatial structure of the molecule and on the interactions with other molecules in the solution, that is, on parameters which cannot be as easily controlled as the chemical structure.

In the case of the ^{1}H-NMR the range of the chemical shifts is comparatively small (Table 2.2) and the effects that depend on the spatial structure and on the interaction with other molecules are so large that the NMR spectrum of a substance cannot be predicted with sufficient accuracy on the basis of the chemical structure.

At room temperature many molecules in solution are not completely rigid but may exist in several different conformations. The equilibrium between the existing conformations depends on the composition of the solvent, on the temperature and on the pressure. Furthermore, the molecules investigated may interact with each other or with other molecules and may form more or less stable complexes. The probability of these interactions depends on the concentration of all components involved. Since the various conformations and complexes may be reflected in the NMR spectrum, the actual spectrum depends in principle on all parameters that influence these equilibria.

In contrast to ^{1}H-NMR a large range of chemical shifts is characteristic for ^{13}C-NMR (Table 2.2) and the direct influence of the chemical structure is predominant. In this case the chemical shift of a substance can be comparatively well predicted on the basis of the NMR spectrum of the basic substance if the effects of the substituents are additively considered as a correction of the chemical shift.

2.1.2 The Internal and External Reference

As shown in Sect. 1.2.1 the chemical shift is usually measured relative to the resonance frequency (the signal) of a reference substance that is either added to the solution being investigated as an internal reference or placed in a tube as an external reference. Hence, the values of the chemical shift δ obtained depend on the resonance frequency of the reference used. Therefore, a value of the chemical shift does not make sense without mentioning the reference substance used. Certain reference molecules have become so generally used that even if the chemical shift is measured with respect to another reference molecule (secondary reference), it is frequently converted for the primary reference.

Internal as well as external reference substances have their advantages and disadvantages: The internal reference can interact with other molecules of the solution which may lead to a shift of the reference signal and thus to a shift of the reference point of the scale. An interaction with the substance investigated may even directly change the spectrum. A necessary precondition for a good internal reference substance is that these effects are to a large extent negligible. It has, of course, to be chemically stable and well soluble in the solvent used. All these conditions are sufficiently fulfilled by tetramethylsilane (TMS), 2,2-dimethyl-2-silapentane-5-sulphonic acid (DSS) and 3-trimethylsilyl-propionic acid (TSP) (Table 2.3), the methyl resonances of which serve as reference point of the ppm scale. They have become generally accepted as internal references in ^{1}H- and ^{13}C-NMR. For other nuclei, like ^{15}N and ^{31}P that play a role in biological NMR, generally accepted internal references do not exist at present.

The external reference is separated in space from the sample investigated; it is mostly placed in a capillary which is immersed in the sample. This has the obvious advantage that it cannot interact with the molecules in the sample. However, the magnetic field at the nucleus does not only depend on the screening effect of the electronic shell of the molecule investigated, but in addition the external magnetic field B_0 is modified by the magnetic susceptibility of the surroundings. In the case of the internal standard this modification of the magnetic field by the susceptibility acts equally on the molecules of the standard and of the sample and thus cancels itself. A problem arises when the solvents of the external standard and of the sample have different magnetic susceptibilities χ. The sample and the standard "see" a different magnetic field which not only depends on the magnetic susceptibilities χ_{sample} of the sample and χ_{ref} of the standard, but also on geometrical factors. In the set-up usually encountered in high resolution NMR where the sample and the reference are placed in coaxial, cylindrical tubes, the long axes of which are parallel to the magnetic field, the theory provides a difference ΔB of the external magnetic fields in reference and sample

Table 2.3. Commonly used references in biological NMR

Spectroscopy	Substance	Comments
^{1}H-NMR	TMS (tetramethyl-silane)	Internal reference substance often used in chemistry. Insoluble in water.
	DSS (2,2-dimethyl-2-silapentane-5-sul-sulfonic acid)	Soluble in water, frequently used in biochemistry. Additional broad signals at 0.60, 1.73 and 2.93 ppm.
	TSP (3-(trimethyl-silyl)-tetradeutero-propionic acid	Frequently used in biochemistry. Soluble in water, chemical shift slightly pH-dependent.
^{13}C-NMR	Same references as for ^{1}H-NMR	See above.
^{15}N-NMR	^{15}NH$_4^+$-ion	No generally accepted "ideal" reference available.
	^{15}NO$_3^-$-ion	Frequently used in combination with the ammonium ion as ^{15}NH$_4$^{15}NO$_3$. ^{15}NO$_3^-$ signal shifted 359 ppm relative to ^{15}NH$_4^+$.
	Nitromethane	354 ppm relative to ^{15}NH$_4^+$.
	Nitrobenzene	349 ppm relative to ^{15}NH$_4^+$.
^{31}P-NMR	Phosphoric acid (85 %)	Most frequently used external reference. Values determined with other reference substances are usually converted to this reference.
	DMMP (dimethyl-methylphosphonate)	Water soluble, can be used as internal reference. 39.4 ppm relative to 85 % phosphoric acid.
	TEP (triethyl-phosphate)	Water soluble, well suited as internal reference. 0.4 ppm relative to 85 % phosphoric acid, chemical shift almost independent of pH.
	Phosphocreatine	Frequently used internal reference for in vivo NMR. Occurring naturally in many tissues. Chemical shift pH-dependent (pK 4.6), at neutral pH – 2.3 ppm relative to 85 % phosphoric acid.

$$\Delta B = -\frac{4\pi}{3}(\chi_{\text{sample}} - \chi_{\text{ref}})B_0 \quad . \tag{2.1}$$

The susceptibility correction factor is especially important if spectra are compared that are measured in superconducting magnets and in conventional electromagnets. Since in the latter the sample is usually placed with its long axis perpendicular to the external magnetic field, the theory gives a factor $+\frac{2}{3}$ instead of $-\frac{4}{3}$ in Eq. (2.1). Consequently, even with identical samples different chemical shifts are obtained without a susceptibility correction factor.

In spherical samples the magnetic field inside and outside is identical. Therefore, the susceptibility problem can partially be avoided by placing the reference substance in a spherical container and then immersing it into the sample.

2.1.3 Multiplet Structure

The second important property which leads to a more precise characterization of the spectrum is the multiplet splitting of the resonance lines by the J-coupling (see Sect. 1.2.3). This J-coupling may be a homonuclear coupling between the spins of identical nuclei or a heteronuclear coupling between the spins of different nuclei. The line splitting shown in Fig. 2.1 originates from the homonuclear coupling between the protons of the various chemical groups in ethanol. Of course, ethanol does not consist only of hydrogen, but also of carbon and oxygen. A heteronuclear coupling with these elements is not observed because their main isotopes, ^{12}C and ^{16}O, do not possess a nuclear spin and therefore do not exhibit a J-coupling. However, a closer look at the spectrum reveals very weak satellite lines on both sides of the main lines, the ^{13}C satellites. They originate from an additional coupling with the rare carbon isotope ^{13}C that is found in natural abundance of about 1% and possesses the nuclear spin $I = \frac{1}{2}$.

The multiplet structures can be completely described by theory; well-functioning simulation programmes exist that calculate the spectra expected on the basis of known coupling constants and chemical shifts. As discussed in Chap. 1, the line pattern in the case of weak J-coupling ($\Delta\delta \gg J$) can be very simply described. Inversely, the number of coupling partners in the chemical structure can be directly derived from the line pattern. If ethanol is mixed with heavy water, the hydroxyl protons are exchanged with deuterons and the signal of the hydroxyl protons in the NMR spectrum vanishes simultaneously with the coupling of the hydroxyl protons to the CH_2-protons (Fig. 2.2). The CH_2- and CH_3-protons are equivalent, respectively; we have therefore an A_2X_3 system. Hence, in agreement with the spectrum (Fig. 2.2) we expect a splitting of the resonance lines of the CH_2-groups into four lines and a splitting of the resonance line of the CH_3-groups into three lines.

The intensities of the absorption lines of the multiplet components follow similarly simple rules as their number. In principle they can be obtained successively by applying the line splittings due to the various couplings on the resonance line one after another and distributing the intensity with each step on the individual

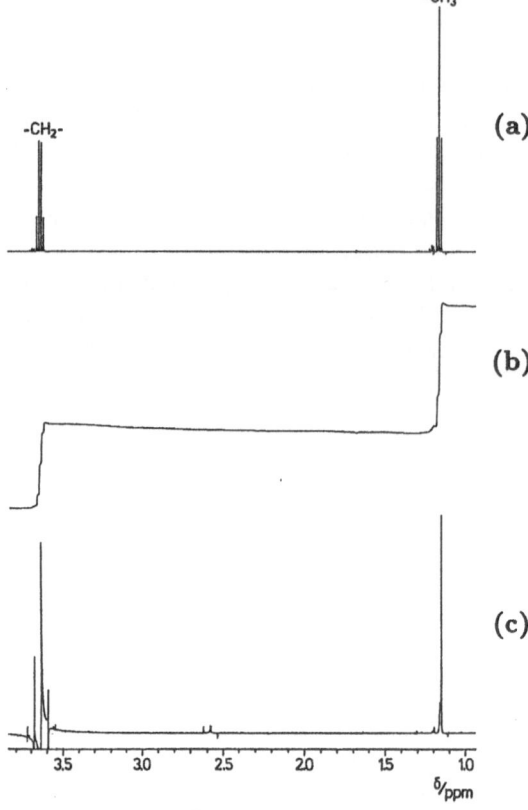

Fig. 2.2a-c. 500 MHz ^{1}H-NMR spectrum of ethanol (concentration: 0.1 %) measured in heavy water with a deuterium content of 99.75 %. The spectrum shown is the Fourier transform of a single FID recorded in 7.2 s. (a) Part of the spectrum showing the methyl and methylene resonance of ethanol. (b) Integral of the spectrum. (c) Decoupling of the methylene resonance at 3.65 ppm leads to a collapse of the triplet at 1.18 ppm

components. For coupling with N equivalent nuclei with the nuclear spin $I = \frac{1}{2}$ that causes a splitting of the resonance line in $N + 1$ multiplet components, the intensities follow the binomial coefficients $\binom{N}{m-1}$. The intensity I_m of the mth component is then

$$I_m \sim \binom{N}{m-1} = \frac{N!}{(m-1)!(N+1-m)!} \, . \tag{2.2}$$

Hence, for the CH$_3$-group in ethanol one obtains for the coupling with the CH$_2$-group ($N = 2$) three multiplet components with an intensity ratio $1 : 2 : 1$.

The case becomes considerably more complicated if the approximation of weak coupling is no longer valid. Figure 2.3 shows the spectrum of the simple AX system that consists of two coupled spins $I = \frac{1}{2}$ if the coupling constant J_{AX}

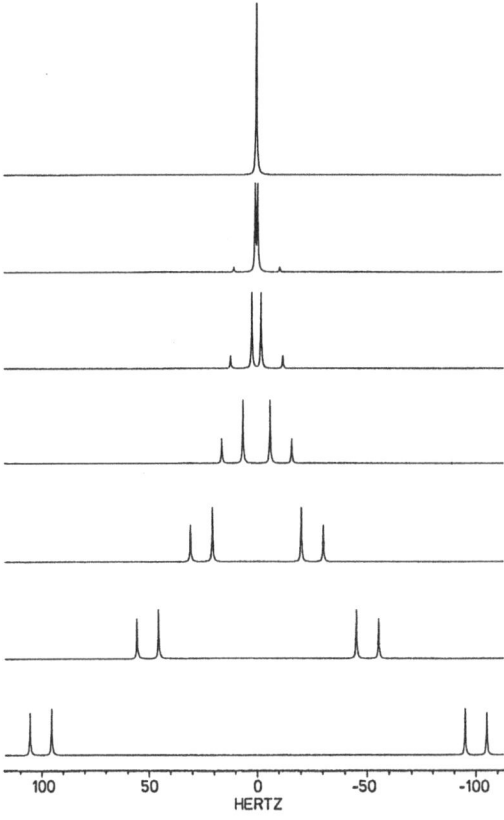

Fig. 2.3. Weak and strong coupling in a two-spin system. Simulation of an AX spin system ($I = \frac{1}{2}$) that changes to an AB spin system by reduction of the chemical shift difference $\Delta\delta$. The scalar coupling constant J_{AX} was kept constant ($J_{AX} = 10\,\text{Hz}$). From top to bottom: $\Delta\delta/J = 0, 0.5, 1, 2, 5, 10, 20$

remains constant, but the difference of the chemical shift $\Delta\delta$ measured in Hz decreases until the weakly coupled system turns into a strongly coupled system. Thus in the nomenclature used the AX system becomes an AB system. The use of letters following each other directly in the alphabet conventionally designates strong coupling. If the chemical shift δ is measured in ppm, as is generally the case, its value is field-independent; because of this property the ppm reference scale is frequently chosen. Hence, the difference $\Delta\delta$ measured in frequency units (Hz) increases in proportion to the external magnetic field, while the coupling constant is field-independent. The strong coupling can therefore be changed into a weak coupling by applying high magnetic fields (Figs. 2.3 and 2.4) to simplify the interpretation of the spectra. This effect complicates, of course, the direct comparison of spectra measured at different magnetic fields.

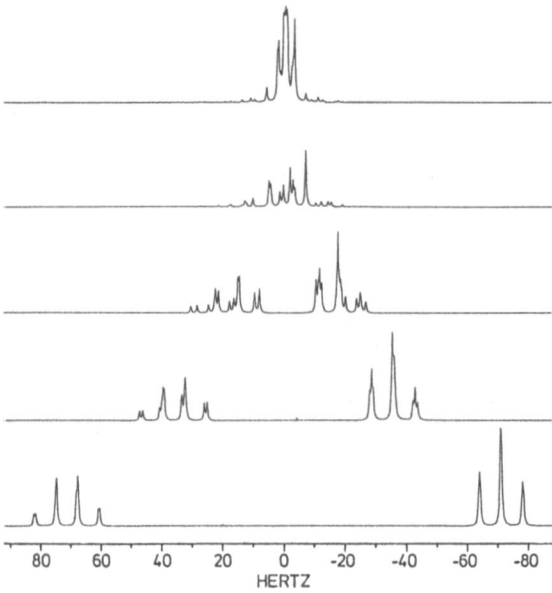

Fig. 2.4. Weak and strong coupling in an A_2X_3 system. The coupling constant J_{AX} of 7.1 Hz used for the simulation corresponds to the one found in ethanol for the coupling between the methyl and methylene protons. Since the chemical shift difference $\Delta\delta$ (measured in frequency units) increases with increasing magnetic field, the spectra show what the NMR spectrum of ethanol would look like at different frequencies: from top to bottom: 3 MHz, 6 MHz, 15 MHz, 30 MHz and 60 MHz spectrum

2.1.4 Structure Dependence of the J-Coupling

The J-coupling between two nuclear spins is brought about by the bonding electrons. Its value decreases to a first approximation with the number of bonds between two nuclei. Hence, in many cases predictions for the structure of an unknown substance can be either rejected or considered to be likely on the basis of the multiplet structure.

On closer inspection, the value of the J-coupling does not depend simply on the number of bonds between the nuclei, but on the details of the electronic distribution in the molecule. The value of J depends, among other things, also on the dihedral angle θ by which two neighboring J-coupled groups are twisted with respect to one another. Due to this, the coupling between two nuclear spins may vanish under unfavorable conditions even if there are just a few bonds between them. As will be seen later, this angular dependence of the scalar coupling provides information in addition to just the primary chemical structure that may play a role when determining protein structures by NMR.

2.1.5 Identification of Coupling Partners

Knowing which nuclei in a molecule are scalar coupled gives valuable information as to its chemical structure, in particular, if the nuclei concerned are neighbors in the chemical structure.

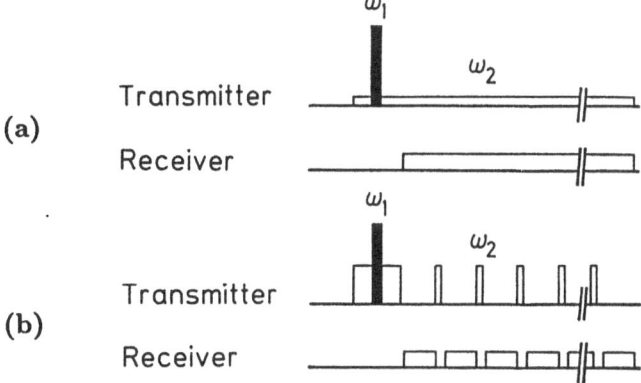

Fig. 2.5a,b. The decoupling experiment. (a) cw decoupling, (b) time shared decoupling. The amplitude of the observation pulse at the frequency ω_1 is usually much higher than that of the decoupling pulse at the frequency ω_2. From theory the decoupling field has to be activated only simultaneously with the ω_1 field; in practice it is activated just before the observation pulse to compensate for electronic imperfections. Relative to the ω_1 pulse the receiver is gated with some delay, so that the ω_1 pulse can decay completely

In our simple case, the spectrum of ethanol, it can be deduced from the coupling pattern that the CH_2 and the CH_3 groups are neighbors to each other, a statement which is obviously correct. In a complex NMR spectrum that contains more than two lines it is often not so easy to decide which coupling to which nuclear spin causes the given multiplet structure. In this case a simple double resonance experiment, called spin decoupling, may be helpful. Selective radio frequency irradiation on the resonance line A while detecting the FID causes the J-coupling effect on the coupled resonance line X to vanish. With ethanol the decoupling of the methylene resonance (Fig. 2.2c) leads to a collapse of the triplet of the methyl resonance. Fig. 2.5 shows schematically the spin decoupling experiment. Simultaneously with the detection pulse of the radio frequency field B_1 a second radio frequency field B_2 at the frequency ω_2 is switched on in addition and irradiated during the total data acquisition time. For an effective spin decoupling the field strength B_2 of the decoupling field has to satisfy the condition $\gamma B_2 \gg 2\pi J$. The strength of this field is much larger than the radio frequency field produced by the excited nuclear spins. In the heteronuclear case this does not cause any experimental difficulties since the spin decoupling frequency and the detection frequency are very far apart. In the homonuclear case the detection of the much weaker NMR signals becomes impossible when irradiating continuously with the much stronger spin decoupling B_2-field. For this reason the ω_2 frequency is not irradiated continuously, but the B_2-field is pulsed with the same pulse program used for the data acquisition (Fig. 2.5). During the pulse the receiver channel is gated off. Data acquisition occurs during the pulse-free intervals (time sharing) and therefore remains essentially undisturbed although the field strength of B_2 averaged over the whole pulse cycle satisfies the condition mentioned above.

The decoupling experiment has nowadays lost much of its importance. As can be seen later, the same information for all coupling partners may be obtained by one two-dimensional experiment that is, however, more time consuming. When studying larger molecules, where correspondingly many individual spin decouplings would be necessary, the longer time needed for a 2D experiment is easily compensated. There exist a number of variations of these double resonance experiments that can provide further information in special cases, but they are of less general interest. One of these cases is spin tickling, which does not simplify the spectrum but causes additional lines and is therefore mostly an unwanted artifact that occurs if the high radio frequency power of the decoupling experiment is not sufficient.

In the heteronuclear case the coupling information is frequently less interesting than the simplification of the spectrum by the collapse of all heteronuclear multiplet splittings. For this purpose not only one but if possible all frequencies have to be selectively decoupled simultaneously. For this broadband decoupling there exist various different experimental solutions. A simple possibility is a fast modulation of the radio frequency used with a standard procedure, the noise decoupling. The more recently developed multipulse techniques (WALTZ, MLEV, GARP) that directly take the tranformation properties of the nuclear spins into account are more efficient and are able to decouple a larger frequency range.

A problem with all broadband decouplings is the fact that radio frequency is irradiated for a relatively long time. In particular in aqueous solutions containing electrolytes this radio frequency may be strongly absorbed and cause heating of the sample. This increase in temperature can cause *in vitro* alterations of the NMR spectrum, or can lead to an irreversible denaturation of sensitive biological molecules. With *in vivo* investigations it may potentially lead to a destructive heating of tissue that cannot be accepted.

2.1.6 Determination of Concentrations by NMR

The determination of concentrations by NMR is very easy under ideal conditions. Since the area below the NMR absorption signal is proportional to the number of nuclear spins involved, it is also proportional to the number of nuclei. Hence, the relative concentration of molecules can be determined directly from the area because the number of molecules in an equal volume is directly proportional to the number of nuclei contained in these molecules; the absolute concentration can also be measured with suitable calibration.

Let us look again at Fig. 2.1: The integrals (the areas) below the individual resonance lines differ considerably. The relative areas below the hydroxyl, methylene and methyl resonance lines are proportional to the number of protons involved (1 : 2 : 3). The same is true when determining the relative concentration for a mixture of different substances. To measure the concentration it is not even necessary to assign all resonance lines to groups of nuclear spins, but one resonance line per molecule suffices, which should be in the range of the spectrum that is free from superpositions.

Although these relative measurements are in principle rather simple, there are several factors that reduce the precision of the procedure. Most of the sources of these errors can be eliminated by simple precautions. Since the determination of concentration plays an important role, in particular in in vivo NMR, some of the sources of these errors will be discussed in the following.

As we have seen in Sect. 1.4.2, the intensity of the resonance lines may be strongly influenced by saturation effects. Since the longitudinal relaxation time T_1 can vary strongly for different groups of nuclear spins, the degree of saturation can vary, and in the extreme case whole resonance lines can vanish if the repetition rate is made too high or the pulse angle too large. Consequently, if the sample contains nuclear spins with different T_1 times, the relative signal intensity of which is to be measured, one has to choose between maximum sensitivity and maximum linearity of the intensity. The maximum sensitivity is determined by the Ernst angle (see Sect. 1.4.2), the maximum linearity is reached by keeping the intervals between the pulse sequences sufficiently long to allow the nuclear spins with the longest relaxation time T_1^{max} to reach thermal equilibrium before the next pulse sequence begins. For this purpose the interval between the pulse sequences should result in a repetition time T_R that is considerably longer than the longest relaxation time T_1^{max}, that is, $T_R \geq 3T_1^{max}$.

As the excitation pulse has a definite width, the corresponding frequency spectrum is not equally intense at all frequencies. If the duration of the pulse is too long, not all resonance lines are equally excited, thus causing differences in the intensity that depend on the frequency difference of the resonance line and the central frequency of the radio frequency pulse.

In order to increase the signal-to-noise ratio, the radio frequency signal is electronically filtered (usually at the stage of the intermediate frequency) in such a way that only the frequency range of interest is passed through. Since, however, an exactly rectangular analogue filter does not exist, the signals at the edge of the spectrum are already somewhat attenuated.

The analogue signal that in principle may assume any value is converted by the analogue-to-digital converter (ADC) into discrete integer numbers. Determination of the area below the resonance line becomes inaccurate if the number of points on the frequency axis is not sufficient to characterize the absorption line. An exact determination of the concentration is not possible if the digital filtering of the data falsifies the areas below the resonance lines, if the noise is too high or if the superposition with other signals renders an exact integration difficult. If the input signal to the ADC is very weak, it can only be converted in a few rough steps because only a small number of values are available for its description, in the limiting case one value only, the zero. Therefore the integral is determined very inaccurately as well. This can be avoided by amplifying the signal so that a sufficient number of values is available. However, it becomes problematic if a very weak signal is to be measured simultaneously with a strong signal. In this case the maximum possible amplification of the input signal is limited by the dynamic range of the ADC because with Fourier spectroscopy the total signal has to be recorded without distortion since every point in the FID influences the whole spectrum; that is, the largest signal must still fit into the dynamic range of the ADC. At present 16-bit (sometimes still 12-bit) ADC are standard in high field spectrometers, that is, the strongest signal should not be $2^{16} = 65536$ (or $2^{12} = 4096$) times more intense than the weakest signal to be detected. Although these figures appear to be rather large, this happens quite often. If, for instance, the ^{1}H-NMR spectrum of a millimolar solution of a substance in water is to be measured, the water peak is already 110 000 times more intense than the signal of the solute.

Considering all sources of errors appropriately, the relative concentration can be determined with an accuracy of a few percent under favorable conditions. If a standard with known concentration is added, absolute concentrations can be measured with the same accuracy. If for experimental reasons it is not possible to use an internal standard for calibration, for instance with in vivo NMR, determination of the absolute concentration becomes comparatively inaccurate. In this case the resonance intensity depends on many external factors, in particular, on the exact position of the sample in the receiving system and on the tuning of the receiving circuit. Hence, the accuracy of these measurements should be regarded critically.

2.1.7 Suppression of Strong Solvent Signals

A frequently occuring problem is the detection of weak signals superimposed on an intense solvent signal because the concentration of the solvent is always much larger than that of the solute. The simplest method to avoid these difficulties is to choose a solvent that does not contain the nuclei to be measured. This is usually done, for instance, with ^{31}P-NMR because most of the common solvents do not contain phosphorus. On the other hand, most of the common solvents contain protons that produce ^{1}H signals. A possibility to avoid this is to deuterate the solvent, which, however, is not possible in all cases. D_2O is a deuterated solvent frequently used instead of normal water. But other common solvents are nowadays also commercially available in the perdeuterated form.

If one has to use a protonated solvent or if the proton signal of the remaining protons even in a highly deuterated solvent is still too strong, it has to be suppressed by spectroscopic methods. The many methods developed for this purpose can be roughly divided into two categories: Either the signal of the solvent is selectively prepared in such a manner that it is strongly attenuated in the following non-selective excitation or the total spectrum is excited but the excitation of the frequency of the solvent is selectively excluded. Of course, both methods may be combined in a suitable manner.

The most common method is the selective presaturation of the resonance line of the solvent with a long weak pulse (Fig. 2.6). The following non-selective detection pulse excites the total spectrum, but the intensity of the solvent peak is attenuated by a factor between 100 and 1000 by the presaturation. Another method is the inversion of the solvent signal by a 180° pulse and excitation

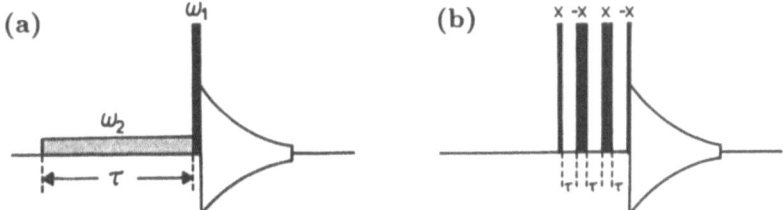

Fig. 2.6a,b. Methods for suppressing strong solvent signals. Two typical methods are shown schematically, selective presaturation (a) and selective excitation with a $1\bar{3}3\bar{1}$ pulse (b). The diagrams are not to scale. The duration τ of the selective presaturation pulse with the frequency ω_2 is comparable to the relaxation time T_1 of the solvent (typical τ for water is 1 s), the observation pulse ω_1 is very short (a few μs). The distance between the pulses in (b) determines the maximum of the excitation that is located in the distance $\Delta\nu = 1/(2\tau)$ on both sides of ω_2. For a maximum excitation at a distance of 1000 Hz one has to select a τ of 0.5 ms. (Since the pulses have a finite length, one has in principle to take their duration into account). The pulse lengths of the individual pulses have the ratio $1:3:3:1$, their phase changes between $0°$ and $180°$ (x and $-x$)

of the total spectrum after a time τ which is chosen in such a way that the magnetization of the solvent is zero at this time.

A disadvantage of all these methods is that other parts of the NMR spectrum may be influenced as well, even if they are not in the immediate neighborhood of the solvent signal. The relevant mechanisms, the saturation transfer and the nuclear Overhauser effect, will be discussed later.

The method of semiselective excitation has the advantage that these effects do not occur. Semiselective excitation pulses can either be produced by long pulses of lower intensity (soft pulses) or by a sequence of pulses of high intensity (hard pulses). The soft pulses can be obtained with almost any frequency spectrum, but they are difficult to produce and to adjust with the standard equipment of a high resolution NMR spectrometer (except with the special hardware which is standard with NMR tomographs). For this reason it is more common to use hard pulses, such as a $1\bar{3}3\bar{1}$ pulse (Fig. 2.6). The principle of these pulse sequences is easy to understand. At a certain frequency the effects of the individual pulses cancel each other, while they do not completely cancel each other for other frequencies and therefore excite these parts of the spectrum. A disadvantage of the semiselective excitation is the usually not easily adjustable phases in the NMR spectrum.

2.2 Time-Dependent Processes in NMR

The nuclear spin relaxation by time-dependent interactions mentioned earlier is an important but by no means the only example of the effect of time-dependent processes in NMR. Such processes can be characterized as being slow or fast with respect to the NMR time scale. The idea behind this is a very useful and simple physical concept that is being intuitively used in many fields. A slow process is best described successively in individual steps, while in the case of a fast process only the time average behaviour is observed. This shows immediately that the concepts "slow" and "fast" only make sense if they are defined with respect to a time scale.

The simplest procedure is to indicate limits beyond which a physical phenomenon can reasonably be described as "slow" or "fast". Obviously these limits will depend on what physical processes are being described in a given approximation. Hence, there exists not only one NMR time scale, but many different scales depending on the interaction concerned. The intermediate time domain in which neither of these two simplifications is valid is most difficult to deal with theoretically. Independent of the NMR time scale, physical motions have their own time scale that can be described by correlation times or characteristic frequencies. The typical frequency ranges for the most important molecular motions are compiled in Table 2.4, and they cover a very wide range from very slow processes of the order of minutes to very fast processes of the order of picoseconds. Almost the whole time domain to about 10 ps can be studied by NMR in some way. Only the very fast processes, like vibrational and torsional motions, can in general not be directly investigated by NMR. In addition, there exist, of course, macroscopic motions, like the flow of liquids or the contraction of muscles. They are of minor importance in the biochemically oriented in vitro NMR, but are more important for NMR tomography and in vivo NMR. This large range of time comprising many orders of magnitude can, of course, not be investigated by one single NMR method, but each NMR experiment is sensitive to motions in a typical time domain.

Table 2.4. Characteristic frequencies of molecular motions

Type of motion	range of frequencies $[s^{-1}]$
Vibrational and torsional motions	$10^{10} - 10^{13}$
Lateral diffusion in membranes	$10^{7} - 10^{10}$
Diffusion in solution	$10^{6} - 10^{13}$
Rotational diffusion in solution	$10^{5} - 10^{12}$
Rotation of aliphatic side chains in macromolecules	$10^{7} - 10^{10}$
Rotation of aromatic side chains in macromolecules	$10^{-2} - 10^{6}$
Conformational changes of proteins	$10^{-5} - 10^{5}$
Release of substrates in enzymatic reactions	$10^{-3} - 10^{6}$

2.2.1 Correlation Time and Spectral Density

Description of molecular motions with a single frequency is a considerable simplification of what really happens. Motions in liquids, as for example the rotation of side chains, do not run evenly over longer periods at a fixed frequency, but the velocity of rotation at a given moment will vary from side chain to side chain and from molecule to molecule and it will frequently vary its magnitude and its direction by collisions with solvent molecules. This means that not only a single frequency but a whole frequency spectrum is obtained.

A simple method to obtain a frequency spectrum from a time modulation was already mentioned several times before, that is, by Fourier transformation. Since there is a large number of similar molecules in solution in different states of motion one averages over the whole ensemble. This leads to a frequency distribution that is usually termed spectral density $J(\omega)$. The precise form of this distribution function depends on the motions observed. The time-dependence of a given quantity $f(t)$ averaged over the ensemble and over a sufficient observation time t is best described by the so-called correlation function $G(\tau)$:

$$G(\tau) = \overline{\langle f^*(t)f(t+\tau)\rangle} \quad . \tag{2.3}$$

The averaging over all times t makes sense if stationary states are described for which it is statistically of no consequence at which time t the correlation function is determined. For stochastic processes, like molecular motions in thermal equilibrium, the correlation function has its maximum value for $\tau = 0$ and decreases with time. For large times τ it must approach zero since in case of stochastic processes the existing correlation must vanish by definition if one waits long enough. The width of the correlation function is usually described by a parameter, the correlation time τ_c, which measures the period during which a quantity can be considered as being correlated.

The spectral density $J(\omega)$ is then the Fourier transform of the correlation function $G(\tau)$

$$J(\omega) = \int_{-\infty}^{\infty} G(\tau)e^{-i\omega\tau}d\tau \quad . \tag{2.4}$$

The precise meaning of the correlation time depends on the corresponding correlation function $G(\tau)$ that is in principle determined by the physical process. However, in many cases the correlation function can be sufficiently approximated by an exponential function

$$G(\tau) = G(0)e^{-(|\tau|/\tau_c)} \quad . \tag{2.5}$$

The corresponding spectral density $J(\omega)$ that is obtained through Fourier transformation of this correlation function is the Lorentzian function mentioned earlier

$$J(\omega) = \int_{-\infty}^{\infty} G(0)e^{-(|\tau|/\tau_c)}e^{-i\omega\tau}d\tau$$

$$= G(0)\frac{2\tau_c}{1+\omega^2\tau_c^2} \quad . \tag{2.6}$$

For the simple case of diffusion the correlation time can easily be interpreted: it is the average lifetime between two collisions with other molecules. The width at half height of the Lorentzian function is $1/\tau_c$. Hence, a decrease of the correlation time leads obviously to an increase of the frequency range at which the spectral density still differs appreciably from zero. The area below the frequency distribution function is constant and independent of the correlation time.

The correlation time τ_c is obviously a suitable quantity to be compared with the NMR time scale discussed above. The time scale itself is determined by the maximum value $\Delta\omega$ of the interaction that causes the phenomenon observed. The amplitude of the interaction $\Delta\omega$ is here measured in frequency units. The process is slow if

$$|\Delta\omega\tau_c| \gg 1 \tag{2.7}$$

is valid, and is fast if

$$|\Delta\omega\tau_c| \ll 1 \tag{2.8}$$

is valid.

2.2.2 Chemical Exchange

Chemical exchange is an important and intuitively easily understandable example of the time-dependent processes discussed above. In the general sense chemical exchange means processes in which the nuclear spin moves to and fro between chemical surroundings that are characterized by different NMR parameters. The change of the surroundings of the nuclear spin can be caused by an intramolecular process, like a change of conformation, or by an intermolecular process. In the latter case the nuclear spin investigated can, for instance, become part of a new covalent structure, or its surroundings may be changed by intermolecular interactions of the molecule concerned. In the simplest case there are only two different states which differ by the chemical shift δ. In this case the correlation time τ_c corresponds to the average lifetimes τ_A and τ_B in the states A and B, and $\Delta\omega$ is just the difference of the chemical shifts $\Delta\delta = \delta_B - \delta_A$ expressed in units of the angular frequency ω. Slow exchange is then defined by the condition

$$|\Delta\delta\tau_c| \gg 1 \tag{2.9}$$

and fast exchange by the condition

$$|\Delta\delta\tau_c| \ll 1 \quad \text{with} \tag{2.10}$$

$$\tau_c = \frac{1}{1/\tau_A + 1/\tau_B} \; . \tag{2.11}$$

The corresponding NMR spectra can be described with an extension of the Bloch equations (McConnell, 1958). The two limiting cases of the slow and fast exchange can be easily understood intuitively without much mathematics by going to the limits $\tau_c = 0$ and $\tau_c = \infty$ respectively. In the first case only a single line

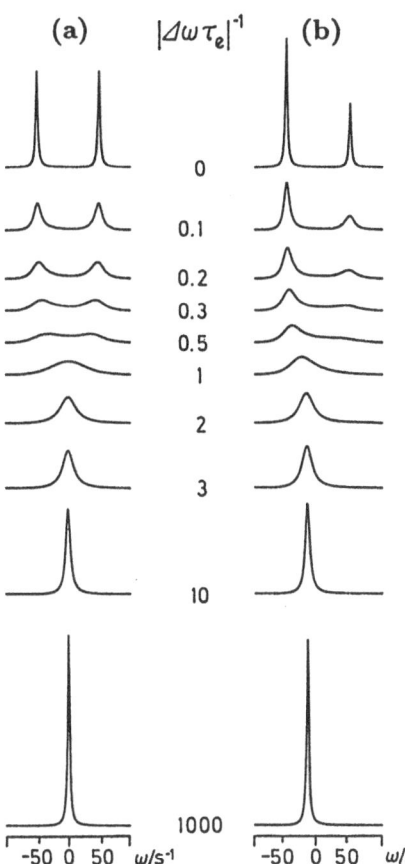

Fig. 2.7a,b. Change of lineshapes with the exchange correlation time τ_e. Simulation of a two-site exchange as a function of $|\Delta\omega\tau_e|$ with an equilibrium constant $K = p_A/p_B$ of 1 (a) and 2 (b) using the McConnell equations

appears because the nucleus exchanges extremely fast with respect to the NMR time scale between the two surroundings. In the second case two resonance lines δ_A and δ_B appear, the intensities of which are proportional to the relative populations (concentrations) p_A and p_B because there is no exchange at all. The averaging of the resonance line positions in the case of fast exchange leads to the chemical shift δ

$$\delta = p_A\delta_A + p_B\delta_B \quad . \tag{2.12}$$

The behavior in the intermediate range does not follow a simple relation, but the NMR signal as a function of the exchange correlation time τ_e has to be calculated explicitly (Fig. 2.7). Qualitatively, starting from the slow exchange the lines become wider and wider until they finally coalesce to one single line if $|\Delta\delta\tau_e| \approx 1$ is satisfied. With increasing exchange rate the line then again becomes narrower until it finally reaches a line width that corresponds to the average value of the line widths without exchange.

A similar averaging process is also found for other interactions. If the two surroundings differ only in their spin-spin coupling constant J, again a multiplet with an average coupling constant $J' = p_A J_A + p_B J_B$ is observed if the relation $|2\pi \Delta J \tau_e| \ll 1$ is valid for the coupling constants measured in frequency units. In many cases the coupling constant J defines a different time scale from the chemical shift, because these two quantities are not primarily correlated. Chemical exchange in J-coupled systems can no longer be described quantitatively with the simple McConnell equations, and has in principle to be treated using quantum mechanical methods.

What sort of biochemical information can be deduced from these processes? In the case of the fast exchange one can in principle calculate the populations of the individual states if there is some additional information available. The most important additional information is certainly the number of states that in fact occur in the fast exchange because the number of states is, of course, relevant for the calculation. For a fast exchange between N states (2.12) has to be extended and averaged over all N states.

$$\delta = \sum_{i=1}^{N} p_i \delta_i \quad . \tag{2.13}$$

Without knowing the δ_i's (2.13) is not defined even for the exchange between two sites. For instance, in the case of a simple chemical equilibrium $A + B \leftrightarrows AB$ the chemical shift of A must be determined in the absence ($p_1 = 1$, $p_2 = 0$) and with an excess of $B(p_1 = 0$, $p_2 = 1)$. In the general case some additional knowledge (models) concerning the functional context of the p_i's is required for the calculations. If it is found that in the particular case studied the NMR spectrum possesses all the properties of fast exchange, a second condition follows immediately: The exchange correlation time τ_e has to be short as compared to $1/|\Delta \delta|$.

In the case of slow exchange one can obtain even more information: the minimum number of states can be counted from the number of resonance lines, and the corresponding populations are directly given by the integral over the resonance lines. This provides the possibility to directly detect different states by NMR which otherwise can only be inferred indirectly, for instance, with kinetic methods. In this case an estimation of the exchange correlation time τ_e can again be given: it has to be larger compared to the inverse of the frequency difference $1/|\Delta \delta|$. In the intermediate range close to $|\Delta \delta \tau_e| = 1$ the exchange strongly influences the lineshape. Hence, the exchange correlation times τ_X in the different states X can be relatively precisely determined by simulation of the resonance lines. For the moderately slow exchange a simple relation results between the observed transverse relaxation time T_{2X}' (and the corresponding line width $1/T_{2X}'$), the relaxation time T_{2X} without exchange ($\tau_X = \infty$) and the exchange correlation time τ_X in the state X:

$$\frac{1}{T_{2X}'} = \frac{1}{T_{2X}} + \frac{1}{\tau_X} \quad . \tag{2.14}$$

Also in the case of slow exchange conclusions have to be drawn cautiously since

NMR gives only a statement on the minimum number of possible states of the system. There are many reasons why possible additional states do not appear in the spectrum. The additional resonance lines may, for instance, be very much broadened, they may be superposed by other lines or be undetectable because there is fast exchange between part of the states in the system. It is clear that conclusions based on an incorrect model are, in general, not correct.

2.2.3 pH-Dependence of the Chemical Shift

Many molecules of importance in biology contain functional groups which may exist in aqueous solution either protonated or deprotonated. In general, an equilibrium between the different forms of these molecules that exists in solution is determined by the concentration of the hydronium ions. The chemical equilibrium can be described in accordance with the law of mass action by an equilibrium constant K that determines the ratio of the concentrations of the molecular species in equilibrium. This simple acid-base equilibrium of the acid AH is usually described in its logarithmic form, the well-known Henderson-Hasselbalch equation

$$pH = pK + \log \frac{[A^-]}{[AH]} \quad .$$ (2.15)

The various charged states of the molecule lead to differences in the electronic distribution and thus in general to a considerable change between the chemical shifts δ_{A^-} and δ_{AH} of the corresponding molecular species that does not concern the functional group itself only, but that can frequently still be detected several bonds away.

In most cases protonating and deprotonating reactions are fast with respect to the NMR time scale; this means that only an average chemical shift δ can be seen, the quantity of which depends on the pH value. Equation (2.15) together with (2.12) then leads to the modified Henderson-Hasselbalch equation

$$\delta = \delta_{AH} + (\delta_{A^-} - \delta_{AH}) \frac{10^{pH-pK}}{1 + 10^{pH-pK}} \quad .$$ (2.16)

If the chemical shift is measured as a function of the pH value and (2.16) adapted to the experimental results, the pK value of individual groups can be determined in a simple manner by NMR. This plays an important role, in particular for proton resonance of proteins where the pK value of individual charged groups can be precisely determined with this method.

Inorganic phosphate participates in many enzymatic reactions, and is either transferred to substrates or removed. Furthermore, it has an important buffer function. For the phosphate group there are not only two different forms of protonation, but in equilibrium there exist H_3PO_4, $H_2PO_4^-$, HPO_4^{2-} and PO_4^{3-}. If the ^{31}P-NMR spectrum of inorganic phosphate is measured at different pH values, only one single absorption signal is observed, the chemical shift of which varies as a function of the hydronium ion concentration (Fig. 2.8). This is precisely

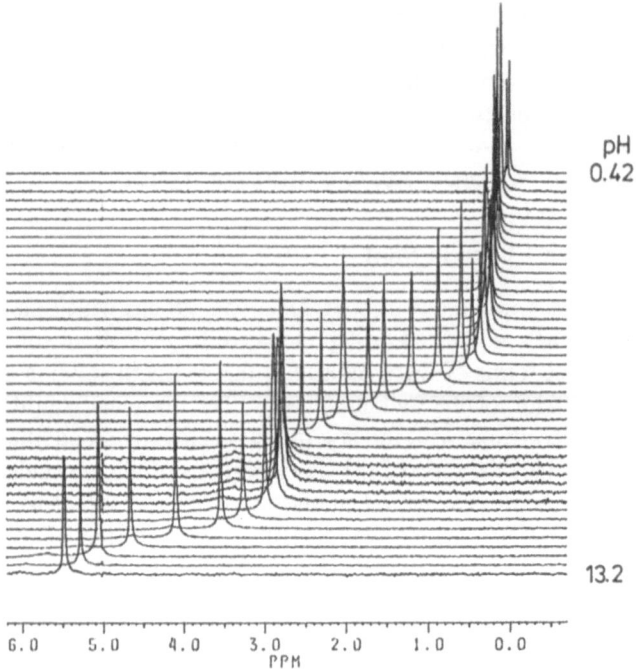

pH
0.42

13.2

6.0   5.0   4.0   3.0   2.0   1.0   0.0
PPM

Fig. 2.8. ^{31}P-NMR spectrum of inorganic phosphate at various pH values. The phosphorus resonance spectrum of inorganic phosphate P_i in a 10 mM phosphate buffer was recorded as a function of pH and plotted sequentially beginning with pH 13.2. The pH value decreases by approximately 0.3 units from spectrum to spectrum

the behavior expected for fast exchange. If the chemical shift δ is plotted as a function of the pH value four titration steps are obtained. This shows directly that at least four different forms with the chemical shifts δ_1, δ_2, δ_3 and δ_4 undergo fast exchange with each other, in complete agreement with the four protonation states of the inorganic phosphate predicted from chemical considerations.

The functional connection is in this case somewhat more complicated than given by Eq. (2.16) because here more than one pK value is required to describe the equilibrium. Applying the law of mass action leads to an extension of (2.16) to N pK values pK_i. If there are no cooperative effects the expression for N pK values can be simply written as the sum

$$\delta = \delta_1 + \sum_{i=1}^{N}(\delta_{i+1} - \delta_i)\frac{10^{\mathrm{pH}-\mathrm{p}K_i}}{1 + 10^{\mathrm{pH}-\mathrm{p}K_i}} \quad . \tag{2.17}$$

If the pK values are as far apart from each other as in the case of the inorganic phosphate, the chemical shifts close to the individual pK values can be described to a good approximation with the simple equation (2.16), a procedure that is frequently applied if only one part of the titration curve is measured.

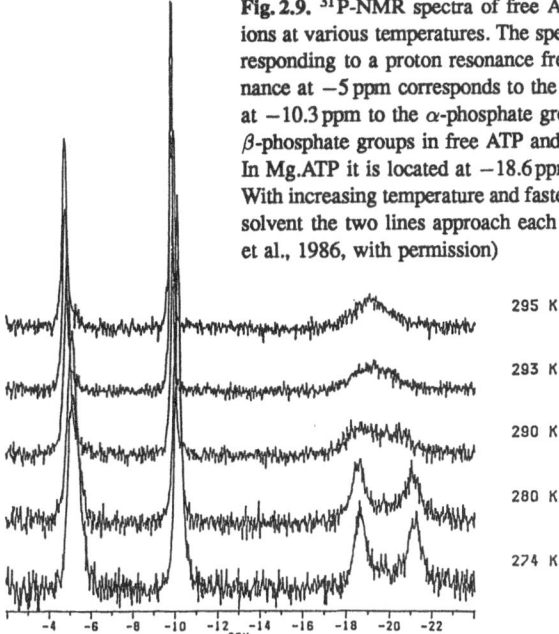

Fig. 2.9. ^{31}P-NMR spectra of free ATP and ATP complexed with Mg^{2+} ions at various temperatures. The spectra were recorded at 203 MHz corresponding to a proton resonance frequency of 500 MHz. The ^{31}P resonance at -5 ppm corresponds to the γ-phosphate group of ATP, the one at -10.3 ppm to the α-phosphate group. At 274 K the resonances of the β-phosphate groups in free ATP and Mg.ATP are separately observable. In Mg.ATP it is located at -18.6 ppm, in metal-free ATP at -21.2 ppm. With increasing temperature and faster exchange of the Mg^{2+} ion with the solvent the two lines approach each other until they merge (Sontheimer et al., 1986, with permission)

295 K

293 K

290 K

280 K

274 K

2.2.4 Formation of Complexes with Diamagnetic Ligands

The formation of complexes of molecules with diamagnetic ligands follows the same rules as the pH-dependent protonation. However, exchange rates are frequently so slow that the condition (2.10) for the fast exchange is not valid anymore. Important for biology are complexes with metal ions which play a role in many enzyme catalyzed processes. In particular, complexes with the divalent ions magnesium and calcium are found very frequently. An important substance for biological reactions, adenosine-5'-triphosphate (ATP), is under physiological conditions almost always complexed with a Mg^{2+} ion.

The lifetime of the Mg.ATP complex is temperature-dependent. At high magnetic fields (11 T) the transition from slow to fast exchange takes place at approximately 25°C. Hence, it is possible to observe the signals of Mg.ATP and free ATP separately at low temperatures, while at higher temperatures only the averaged resonance line is seen, the line position of which is shifted depending on the concentration (Fig. 2.9).

Lineshape analysis permits a very precise determination of the exchange rates and of the activation energies which, however, depend strongly on the buffer system used in agreement with expectation (Table 2.5). Contrary to this, the lifetime of the calcium complexes is always so short that there is always a fast exchange at the highest magnetic fields available at temperatures above the freezing point of water.

The proper substrate in many enzymatic reactions is Mg.ATP. The enzyme activity depends therefore primarily on the concentration of the free Mg.ATP.

Table 2.5. Activation enthalpies and exchange rates in Mg.ATP complexes*

$\Delta G^{\ddagger}/(kJM^{-1})$	k_{-1}/s^{-1} (1° C)	k_{-1}/s^{-1} (10° C)	k_{-1}/s^{-1} (20° C)	k_{-1}/s^{-1} (30° C)	k_{-1}/s^{-1} (37° C)
41	400	600	1100	1900	2900

* The exchange rates $k_{-1} = 1/\tau_A$ (τ_A lifetime of the Mg.ATP complex) and the activation enthalpy $\Delta G^{\ddagger}$ are strongly influenced by the pH-value of the solution and the ions contained in it (for instance, buffer ions). The values shown here are calculated from the spectra of Fig. 2.9.

This concentration can be measured directly with NMR at low temperatures from the intensity of the corresponding resonance lines. At higher temperatures it is still possible to determine the concentrations indirectly from the chemical shift if the binding constant of ATP for magnesium under the given experimental conditions is known.

2.2.5 Saturation Transfer

Another method to determine the exchange rate in the slow exchange region is saturation transfer. It permits in many cases a more precise and unambiguous determination of the exchange correlation times than the analysis of the lineshape. Furthermore, saturation transfer provides additional information on whether two resonance lines do really represent two states coupled by chemical exchange. The method is in principle very simple: Resonance B is saturated by a frequency selective pulse of a duration τ and directly afterwards a normal NMR spectrum is recorded. The experimental procedure corresponds indeed to the one used for the selective presaturation of strong solvent signals.

If the presaturated spins change their position from B to another position A in the time τ, the line intensity of the corresponding absorption line A is reduced. The size of the effect depends on the one hand on the length of the period τ during which the exchange can take place and on the exchange rate $1/\tau_A$. On the other hand, the approximation to the Boltzmann equilibrium characterized by the longitudinal relaxation time T_{1A} reduces the effect. Formally the intensity I_A of the resonance line A as a function of the time τ during which the resonance B is saturated may be described as

$$I_A(\tau) = \frac{I_A(0)}{1/\tau_A + 1/T_{1A}} \left(\frac{1}{\tau_A} e^{-\tau(1/\tau_A + 1/T_{1A})} + \frac{1}{T_{1A}} \right) \quad . \tag{2.18}$$

For saturation times that are considerably longer than T_{1A} an equilibrium is reached and the ratio of the line intensities with or without saturation is simply given by

$$\frac{I_A(\infty)}{I_A(0)} = \frac{1}{1 + T_{1A}/\tau_A} \quad . \tag{2.19}$$

Another possibility to label spin B is the inversion of the magnetization B with a selective 180° pulse (inversion transfer). The effect on the resonance line A is detected after a delay time τ with a non-selective detection pulse. In this experiment the intensity of the resonance line A first decreases with increasing τ until it reaches a minimum value and then again asymptotically approaches its original value in the unperturbed system.

Sources of errors which can falsify the interpretation of the experiments are the limited selectivity of the inversion and saturation pulses, which may have an unwanted influence on the neighboring resonance lines and the nuclear Overhauser effect that can also exert an indirect influence on the signal intensity of A.

2.2.6 Nuclear Overhauser Effect

The change of the intensity of the nuclear resonance signal when saturating the corresponding electron spin resonance was first treated theoretically by A.W. Overhauser (1953) and was later termed the Overhauser effect after its discoverer. The application of this experiment to a system of dipolar coupled spins is the most important basis for the determination of the spatial structure by NMR. The Overhauser effect and the nuclear Overhauser effect (NOE) originate from polarization changes in coupled systems if the population of one or several subsystems is changed. Since the dipolar coupling that produces the population change depends on the distance of the interacting nuclear spins, the size of the nuclear Overhauser effect depends on the distance as well. This dependence offers the possibility to determine the corresponding interatomic distances.

The experimental procedure for determining the NOE is similar to the one we already know for determining exchange rates in the case of saturation transfer or inversion transfer. However, in this context they are usually called truncated driven Overhauser effects (TOE) or transient NOE. Furthermore, the NOE can be measured by two-dimensional NMR experiments that can also be used to study exchange processes. These methods will be treated later in Sect. 2.3.

A simple system demonstrating the characteristic properties of the nuclear Overhauser effect consists of molecules that contain two dipolar coupled nuclei A and B with the nuclear spins $I^A = I^B = \frac{1}{2}$. Averaging over many such molecules, populations can be assigned to each of the possible four states to determine the probability with which each of these states is populated (Fig. 2.10). In thermal equilibrium a Boltzmann distribution is obtained. If one of these populations is changed by selective perturbation, for instance by an irradiating radio frequency at the resonance frequency ω_B of the nuclear spin B, in principle all populations in the interacting total system are changed because of the dipolar coupling. Finally, after a long time irradiation a new dynamic steady state is reached. The relevant relaxation processes may be described following the change of the quantum number m as zero-quantum ($\Delta m = 0$), one-quantum ($\Delta m = \pm 1$) and two-quantum ($\Delta m = \pm 2$) transitions.

These population changes lead also to a change of the z component M_z^A of the magnetization of the spins A, the size of which at a time t can be detected with a

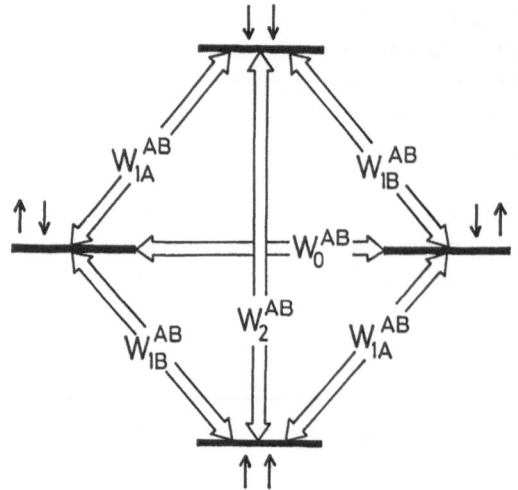

Fig. 2.10. Transition probabilities in a two-spin system. Schematic representation of the different possible orientations of the magnetic moments. W_0, W_1 and W_2 are the transition probabilities for transitions between the different states. The states connected by W_0 usually do not have exactly the same energy as shown here in a simplified form

90° pulse. After Fourier transformation an intensity change of the corresponding resonance line at ω_A is found.

Formally the time-dependent change of the magnetization may be described by rate equations

$$\frac{dM_z^A}{dt} = -\varrho_A(M_z^A - M_0^A) - \sigma_{AB}(M_z^B - M_0^B) \tag{2.20}$$

$$\frac{dM_z^B}{dt} = -\varrho_B(M_z^B - M_0^B) - \sigma_{BA}(M_z^A - M_0^A) \quad . \tag{2.21}$$

In these equations ϱ_A and ϱ_B are the spin-lattice relaxation rates of the nuclear spins A and B, and σ_{AB} is the cross relaxation rate between A and B; M_0 is again the steady state magnetization in thermal equilibrium. The individual relaxation rates are defined by

$$\varrho_A = W_0^{AB} + 2W_{1A}^{AB} + W_2^{AB} + R_{\text{ext}}^A \tag{2.22}$$

$$\varrho_B = W_0^{AB} + 2W_{1B}^{AB} + W_2^{AB} + R_{\text{ext}}^B \tag{2.23}$$

$$\sigma_{AB} = \sigma_{BA} = W_2^{AB} - W_0^{AB} \quad . \tag{2.24}$$

Here W_0^{AB}, W_1^{AB} and W_2^{AB} are the probabilities for zero-, one- and two-quantum transitions and R_{ext} symbolizes additional individual relaxation pathways for spin A and B. In this case the NOE is defined by the signal intensity S after the perturbation divided by the signal intensity S_0 in the unperturbed system or by the quotient M_z/M_0 because of the proportionality of signal intensity and z-magnetization. The radio frequency irradiation at the resonance frequency ω_B over a time τ, which is long compared to the longitudinal relaxation times of the system, with a sufficiently high power for saturating the transition at the frequency ω_B ($\gamma_B^2 B_2^2 T_1 T_2 \gg 1$) causes M_z^B, dM_z^A/dt and dM_z^B/dt in (2.20) and (2.21) to become zero. A simple transformation and the insertion of results for the nuclear Overhauser effect in the steady state gives

$$\text{NOE} = \frac{M_z^A}{M_0^A} = 1 + \frac{\sigma_{AB}}{\varrho_A} \frac{M_0^B}{M_0^A} \quad . \tag{2.25}$$

Since the equilibrium magnetization is proportional to the magnetogyric ratio γ, (2.25) can be rewritten as

$$\text{NOE} = 1 + \frac{\sigma_{AB}}{\varrho_A} \frac{\gamma_B}{\gamma_A} \quad . \tag{2.26}$$

For dipolar relaxation one obtains for the transition probabilities

$$W_0^{AB} = \tfrac{1}{2} q^{AB} J(\omega_0^A - \omega_0^B) \tag{2.27}$$

$$W_{1A}^{AB} = \tfrac{3}{4} q^{AB} J(\omega_0^A) \tag{2.28}$$

$$W_{1B}^{AB} = \tfrac{3}{4} q^{AB} J(\omega_0^A) \tag{2.29}$$

$$W_2^{AB} = 3 q^{AB} J(\omega_0^A + \omega_0^B) \quad . \tag{2.30}$$

If the motion that causes the relaxation is an isotropic random motion like the rotational diffusion of the molecule, then the factor q^{AB} which is proportional to the strength of the interaction is given by

$$q^{AB} = \frac{1}{10} \left(\frac{\mu_0}{4\pi}\right)^2 \frac{\hbar^2 \gamma_A^2 \gamma_B^2}{r_{AB}^6} \quad . \tag{2.31}$$

Here μ_0 is the magnetic permeability of the vacuum and r_{AB} the distance between spin A and spin B. The corresponding spectral density function is:

$$J(\omega) = \frac{2\tau_{\text{rot}}^{AB}}{1 + (\omega \tau_{\text{rot}}^{AB})^2} \quad . \tag{2.32}$$

Inserting these functions into the expressions for ϱ and σ leads to a dependence of the steady state nuclear Overhauser effect on the correlation time τ_{rot} as shown in Fig. 2.11 for several nuclei. When calculating these functions the external relaxation was neglected, which can lead to a decrease of the highest observable effects. It is interesting to note that without external relaxation the dependence of the steady state NOE on the distance r_{AB} between the nuclear spins vanishes as well. For short correlation times typical for small molecules in low viscosity solvents one obtains in the limit

$$\text{NOE} = 1 + \frac{1}{2} \frac{\gamma_B}{\gamma_A} \quad . \tag{2.33}$$

If the saturated nuclear spin B is a proton and if the magnetogyric ratio γ_A is positive, an enhancement of the resonance line A is obtained. For ^{13}C according to (2.33) the maximum enhancement of the signal is 2.988 when saturating a proton bound to it. This increase in the sensitivity is an additional reason why in ^{13}C-NMR the protons are frequently broadband decoupled. However, one has to

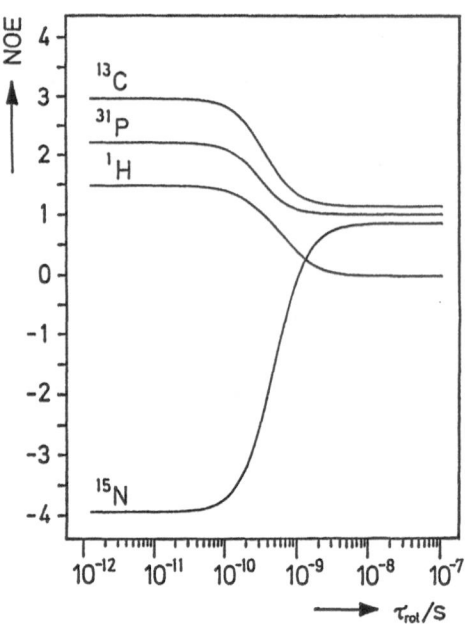

Fig. 2.11. The NOE as a function of the rotational correlation time τ_{rot} at a proton resonance frequency 500 MHz. The NOE was calculated for different nuclei relaxed by a dipolar coupled proton neglecting any external relaxation. A complete long-lasting saturation of the proton resonance was assumed (equilibrium NOE)

be cautious and cannot always expect a signal enhancement due to the NOE. In the case of nuclei, such as ^{15}N and ^{29}Si, that have a negative magnetogyric ratio, the nuclear Overhauser effect may lead to a vanishing of the signal (Fig. 2.11).

If the external relaxation is different from zero, the NOE in the steady state depends on the distance and can hence be used to determine the distance between the nuclear spins concerned. For a quantitative determination of the distance, however, additional information or additional assumptions on the external relaxation of the system are required.

The interatomic distances can be determined more precisely if the time-dependence of the NOEs is observed. Figure 2.12 shows a TOE between a methyl group and an aromatic ring in the HPr protein, a component of the phospho*enol*pyruvate dependent phosphotransferase system. The effect of the saturation pulse at the resonance frequency ω_B can best be quantitatively measured if not only a spectrum with presaturation at the frequency ω_B (on-resonance spectrum) is taken, but also a reference spectrum with a presaturation at a frequency $\omega_{B'}$, at which there is no resonance line in the spectrum (off-resonance spectrum). In this case the difference between the two spectra shows selectively the Overhauser effect. Frequently, the observable effects are of the order of a few percent which means that they are comparatively small. In order to obtain a sufficient signal-to-noise ratio it is usually necessary to average over a large number of individual spectra. Since small instrumental instabilities may falsify the difference measurement, the experiment is usually performed in a cyclic manner by accumulating alternatingly a small number, for instance sixteen on-resonance and off-resonance experiments and recording separately the corresponding FIDs.

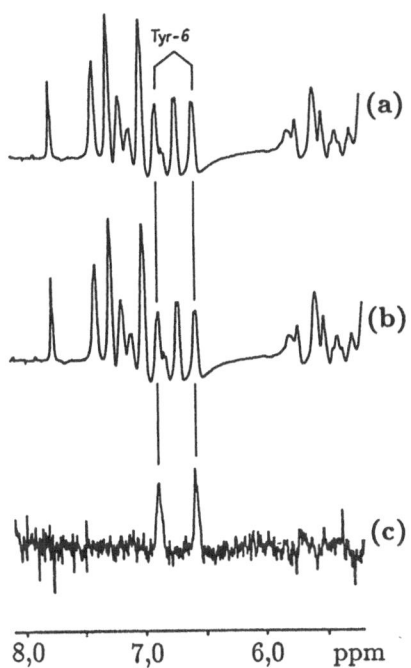

Tyr-6

(a)

(b)

(c)

8,0   7,0   6,0   ppm

Fig. 2.12a-c. NOE difference spectrum of HPr protein. Only the downfield part of the ^{1}H-NMR spectrum (360 MHz) is shown where the aromatic residues have their resonances. TOE experiment with a presaturation of 0.8 s. (a) irradiation off-resonance, (b) saturation of a methyl resonance at -0.18 ppm, (c) difference of (a) and (b) (Rösch et al., 1981a, with permission)

With this method NOEs in the range of several percent can be detected with modern spectrometers.

If the NOEs are plotted as a function of the duration of the presaturation (TOE) or the distance between the 180° and the 90° pulse (transient NOE), the slope at the beginning is determined by the interatomic distance r_{AB} of the nuclear spins concerned: it is proportional to r_{AB}^{-6}. The determination of the distance using the Overhauser effect is an essential part of structure determination of biological macromolecules by NMR. Hence, we shall treat this problem in more detail in Chap. 3.

2.2.7 Labelling with Stable Isotopes

One of the standard methods in biochemistry for following the course of reactions is labelling with radioactive isotopes. Also for NMR it is an important experimental method, however, in this case with stable isotopes. If the course of the chemical reaction is well-known, it is a useful method for assigning resonance lines. If for example a proton in a well-defined position is replaced by a deuteron, the vanishing ^{1}H-NMR resonance line belongs to this position. In order to follow the course of reactions, the naturally occurring isotope can be replaced by a different one so that its course can be continuously followed in the NMR spectrum.

A typical example in this context is the replacement of ^{14}N by ^{15}N, of ^{12}C by ^{13}C and of ^{1}H by ^{2}H. This labelling by isotopes is particularly important for in vivo NMR as we shall see in Chap. 5. Problematic are nuclei like ^{31}P for which no stable isotopes exist that could be used for replacing the naturally occurring isotope. In this case labelling of a whole group is helpful. The course of the reaction of phosphate groups can be followed by replacing ^{16}O with the isotopes ^{17}O or ^{18}O the natural abundance of which is very low.

Labelling with ^{17}O with a nuclear spin of $I = \frac{5}{2}$ leads to a splitting of the phosphorus signal into a sextet. The quadrupole moment of the ^{17}O nuclear spin leads to a fast relaxation; the coupling between the oxygen and the nuclear spin of the phosphorus is an efficient relaxation mechanism that causes additional line broadening of the phosphorus resonance lines. If more than one ^{17}O atom is bound to the phosphorus atom, the distribution of the intensity on many multiplet components causes the signal to vanish in the noise. Labelling with the isotope ^{18}O has a different result: it does not possess a nuclear spin like the abundant isotope ^{16}O and therefore does not lead to a multiplet splitting. However, replacing an ^{16}O atom in the phosphate group by an ^{18}O atom leads to a small highfield shift (isotope shift) of the resonance line of about 0.025 ppm that increases proportional to the number of bound isotopes. This permits one to observe separately the phosphorus resonance signals of the phosphate groups that differ in the number of bound ^{18}O atoms.

The ^{18}O isotope exchange can be used for analysing the course of enzymatic reactions. A large group of enzymes (phosphotransferases) cleave high energy phosphates during the catalysis and transfer a phosphate group either to a different substrate (kinases) or the surrounding water (hydrolases). In the case of the enzymatic hydrolysis of high energy nucleoside triphosphate NTP an intermediate state occurs where the phosphoric acid anhydride bond is already broken, but the phosphate remnant P_i is still bound to the enzyme E together with the nucleoside diphosphate NDP. Formally this can be expressed by the reaction equation

$$\text{E.NTP} + \text{H}_2\text{O} \underset{\rightleftarrows}{^1} \text{E.NPD} \cdot P_i \underset{\rightleftarrows}{^2} \text{E.NPD} + P_i \quad . \tag{2.34}$$

All reaction steps in (2.34) are, of course, in principle reversible, but the speed of these processes depends on the activation enthalpy $\Delta G^{\ddagger}$, that is, on the details of the enzymatic catalysis. For many enzymes step (1) is repeated several times until finally the phosphate group passes to the water. If the terminal phosphate group in NTP is completely labelled with ^{18}O and the reaction takes place in ordinary water, at each step (1) an ^{16}O atom is introduced in the phosphate group. If the reverse reaction (1) does not take place before P_i is split off, only $\text{P}^{16}\text{O}^{18}\text{O}_3$ is found in the solution. In the other extreme case, if step (1) occurs many times before P_i is split off, exclusively free P^{16}O_4 is detected. In general one finds a time-dependent distribution of the isotope labelling that may be simulated with suitable methods. Fitting the data with a suitable model gives the average number of back reactions before the splitting of P_i and the rate constants k_{-2} for the binding of P_i to the E.NDP complex. Figure 2.13 shows such an example for the myosin fragment $S1$. However, in this case instead of using spin labelled ATP the back reaction was observed with isotope labelled inorganic phosphate (P^{18}O_4)

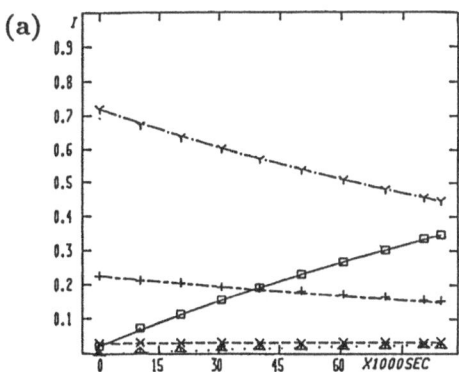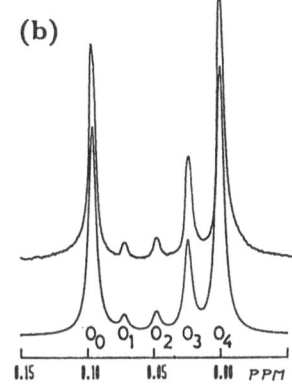

Fig. 2.13a,b. Catalysis of the ^{18}O exchange by the $S1$ fragment of myosin complexed with Mg.ADP (Rösch et al., 1981b, with permission). The ^{31}P-NMR spectrum was recorded at 146 MHz (corresponding to a proton resonance frequency of 360 MHz). (a) Change of concentrations of the individual isotopomers of inorganic phosphate P_i: $H_3P^{18}O_4$ (Y), $H_3P^{18}O_3{}^{16}O_2$ (+), $H_3P^{18}O_2{}^{16}O_2$ (×), $H_3P^{18}O^{16}O_3$ (△) and $H_3P^{16}O_4$ (□). (b) ^{31}P-NMR spectra of the isotopomers of P_i after a reaction time of 23.5 hours (top, experimental spectrum; bottom, simulation of the experimental spectrum)

in the presence of the $S1$.Mg.ADP complex. In this system step (1) is repeated about 40 times before the phosphate is split off. It is worth mentioning in this connection that the first experiments for detecting the enzymatic isotope exchange in phosphate groups were performed with mass spectroscopy that permits one to detect with high sensitivity the isotopic combinations in small molecules. The NMR method has the advantage that the isotopic combination can be measured continuously without derivatisation. However, if low concentrations have to be used, then mass spectroscopy is clearly superior to NMR spectroscopy.

2.3 Two-Dimensional NMR Spectroscopy

The development of two-dimensional NMR has opened new possibilities for the application of NMR in biology. In principle, any series of NMR experiments, where besides the time t_2 or the frequency ω_2 after Fourier tranformation a second parameter is varied, is a two-dimensional NMR experiment. An example of this is a series of one-dimensional NMR spectra that are taken during the course of a chemical reaction at different times t. Plotting these spectra on a second time scale t_1 provides a two-dimensional spectrum. This type of 2D-spectroscopy was, of course, known from the beginning of NMR.

2D-spectroscopy in the proper sense requires in general that the second variable be closely connected with the specific properties of the spin system. At present there are many different methods for producing two-dimensional NMR spectra, which are always plotted as a function of two frequency variables after computer-aided processing. Two-dimensional NMR spectroscopy has several advantages as compared to the one-dimensional method: (1) The information is distributed in two dimensions. It is, therefore, possible to resolve complex spectra that cannot be interpreted in one dimension because of strong superpositions of lines. (2) Simultaneously with this distribution in two dimensions one can select physical interactions which one wishes to separate in these two dimensions. (3) Furthermore, it is relatively simple in two-dimensional spectroscopy to observe indirectly multi-quantum transitions that are spin-forbidden to first approximation.

In principle one can extend this method to multi-dimensional NMR spectroscopy by introducing further time variables; however, the necessary measuring time increases with the number of additional dimensions. For the structure determination of macromolecules three- and four-dimensional methods (3D-NMR and 4D-NMR) in conjunction with isotope labelling appear very promising.

2.3.1 The Two-Dimensional NMR Experiment

Hitherto, only methods using Fourier spectroscopy have been of practical importance in two-dimensional NMR spectroscopy. With these methods a time signal $s(t_1, t_2)$ is detected that is then transformed by Fourier transformation into a two-dimensional NMR spectrum $S(\omega_1, \omega_2)$ in the frequency domain. In principle it is also possible to obtain two-dimensional NMR spectra with an alternative method, stochastic NMR, which is, however, still being developed.

The basic experiment of two-dimensional NMR in the time domain can be divided schematically into four periods (Fig. 2.14): the preparation, evolution, mixing and detection phase. During the detection phase the signals are detected, digitized and stored as in the case of one-dimensional NMR. The preparation phase consists usually of a simple 90° pulse that produces the transverse magnetization. During the evolution phase with a variable time t_1 coherences develop that are coupled to each other during the mixing phase, which is not required in

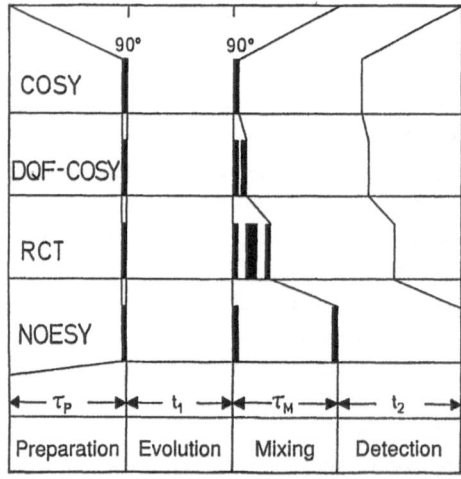

Fig. 2.14. Schematic representation of some important two-dimensional NMR experiments. From the different phases shown, only the mixing phases differ that consist of a single 90° pulse in the COSY experiment, of two 90° pulses shifted by 90° in the double-quantum filtered COSY, of three pulses (90°-180°-90°) in the RCT experiment and of two pulses separated by a rather long time period in the NOESY experiment. Not shown are the complicated phase cycles that are used for suppressing artifacts

some special cases, and is then transformed into a detectable transverse magnetization. From experiment to experiment the time t_1 is increased by a constant value Δt_1, the size of which is determined by the Nyquist theorem, as it is in the t_2 dimension. Each spectrum belonging to a given t_1 value is stored separately. Thus a two-dimensional matrix is obtained that assigns to each pair (t_1, t_2) a signal amplitude $s(t_1, t_2)$. The time signal $s(t_1, t_2)$ is then transformed by a two-dimensional Fourier transformation into a frequency signal $S(\omega_1, \omega_2)$. The complex 2D Fourier transformation can be written as

$$S(\omega_1, \omega_2) = \int_{-\infty}^{\infty} \int_{-\infty}^{\infty} s(t_1, t_2) e^{-i\omega_1 t_1} e^{-i\omega_2 t_2} \, dt_2 dt_1 \qquad (2.35)$$

In fact, (2.35) means that first all FIDs are Fourier transformed as a function of t_2, as usual. The data matrix obtained in this manner contains the NMR spectrum in the order of the t_1 values in the rows of the ω_2 direction. Next, one Fourier transforms once more in the t_1 direction, that is, the data of the individual columns are again considered as FIDs and Fourier transformed as usual.

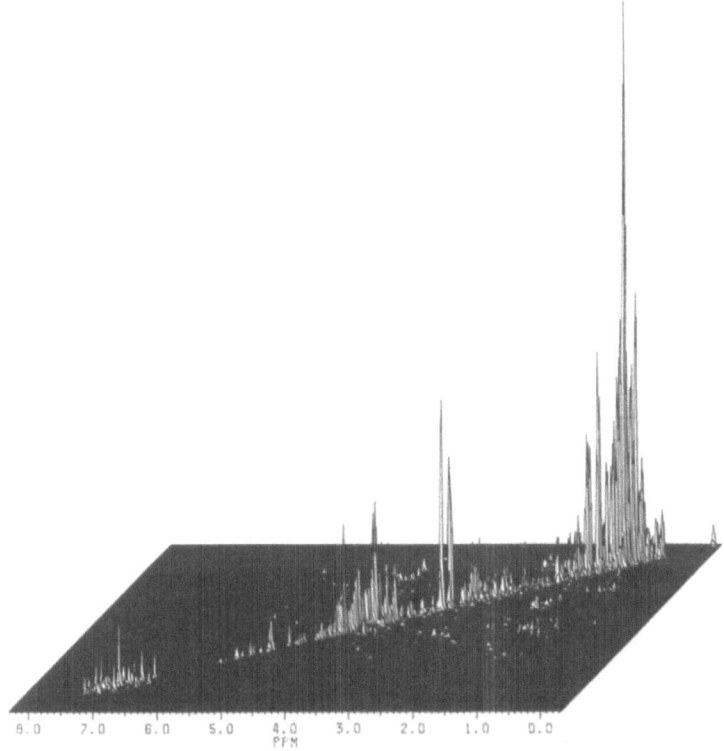

8.0   7.0   6.0   5.0   4.0   3.0   2.0   1.0   0.0
PPM

Fig. 2.15. Stacked plot of a double-quantum filtered COSY spectrum of HPr protein from *Staphylococcus aureus*. Every second spectrum of the data matrix consisting of 512 × 512 points is depicted sequentially. Representation as absolute value spectrum, that is, only the absolute value of the signals having positive and negative parts are shown

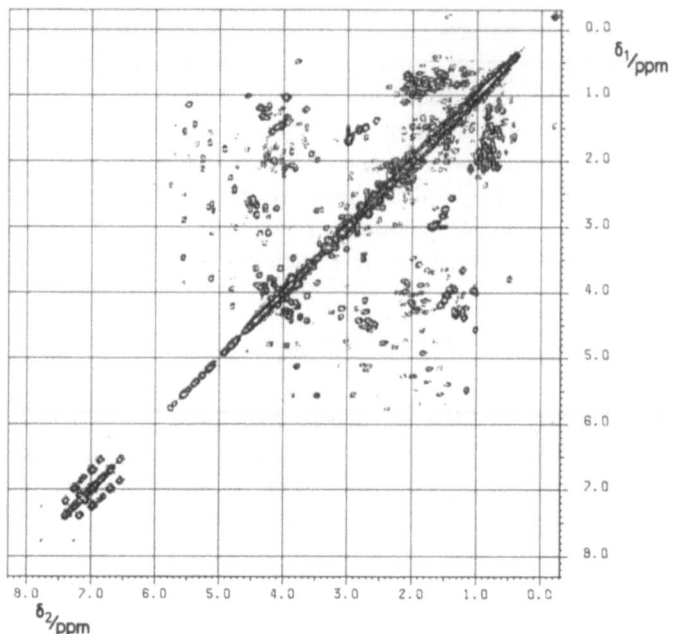

Fig. 2.16. Contour plot of the two-dimensional NMR spectrum of HPr protein. The contour lines were calculated and plotted from the data of Fig. 2.15

As in the one-dimensional case, the Fourier transformation leads to a real and an imaginary part. In general, however, only the real part or the absolute value of $S(\omega_1, \omega_2)$ is plotted. The intensities of the two-dimensional spectrum span a surface in space in the third dimension comparable to the surface of the earth in a plot of a landscape. Two types of plots are now generally accepted for the graphic representation of two-dimensional spectra: the stacked plot and the contour plot. The stacked plot corresponds to a three-dimensional presentation of the surface and provides an illustration of the 2D-spectrum (Fig. 2.15). In the case of the contour plot the surface is viewed perpendicularly from above like looking at a map. Planes of equal signal intensity are represented by contour lines (Fig. 2.16). This type of plot is mostly used for evaluating the 2D-spectra since it presents the geometrical relations without distorsions and since weak peaks are not be covered by strong signals.

The individual steps that lead to a two-dimensional NMR spectrum are compiled again in Table 2.6. If one considers all variants, there exist so far several hundred different types of 2D-spectroscopy. Fortunately, only a few play a role in biological applications. One can distinguish between homonuclear and heteronuclear experiments. In homonuclear experiments only properties of like spins are investigated, for instance of protons; the 2D-pulse sequence consists only of pulses in the corresponding frequency range. In the case of heteronuclear experiments interactions between different nuclear spins, for instance ^{13}C and ^{1}H are observed; the corresponding pulse sequences contain pulses of each frequency range.

Table 2.6. Processing of 2D data *

1.	Arrangement of the FIDs at different time increments t_1 in a matrix $s(t_1, t_2)$. The N_1 rows of the matrix contain the single free induction decays (N_2 data points) arranged according to increasing t_1.
2.	Digital filtering of all FIDs: multiplication of every row with a suitable filter function.
3.	Fourier transformation of the filtered FIDs; the rows of the matrix now contain the spectra $s'(t_1, \omega_2)$.
4.	Phase correction of the spectra $s'(t_1, \omega_2)$.
5.	Digital filtering in the t_1 direction, that is, multiplication of the columns with a suitable filter function.
6.	Completing the matrix by adding rows containing zeros until the digital resolution in both dimensions are equal.
7.	Fourier transformation of the columns.
8.	Phase correction in the t_1 direction.
9.	Representation of the 2D spectrum $S(\omega_1, \omega_2)$.

* Depending on the type of experiment, processing can differ slightly from this scheme.

The biological applications of 2D-spectroscopy make use mainly of two interactions between nuclear spins, the scalar coupling and the dipolar coupling.

2.3.2 Interpretation of Homonuclear J-Coupling Patterns

As explained earlier, the J-coupling is closely related to the covalent chemical structure. In the case of known chemical structures it may serve to assign the resonance lines. In the case of unknown structures different structural hypotheses may be tested. In general two questions have to be answered: (1) which nuclear spins are coupled with each other in the structure and (2) what is the magnitude of this coupling. Both questions can be answered in simple cases by applying one-dimensional methods already discussed, that is using decoupling experiments and simulating multiplet structures. In more complex superposed spectra these methods will succeed, if at all, only after many long and painstaking experiments. Applying two-dimensional spectroscopy this information is obtained by one single experiment. Here the standard experiment is the COSY-experiment (correlated spectroscopy) that consists of two 90° pulses separated by the evolution time t_1 (Fig. 2.14). The spectrum obtained is symmetric to the diagonal on which the so-called diagonal peaks are situated. With its information content the diagonal corresponds essentially to the one-dimensional spectrum. Substantial information comes from the off-diagonal peaks, the cross peaks (Figs. 2.15 and 2.16). They give information on which nuclear spins have J-couplings to each other that are not much less than the line widths of the multiplet components. In principle, the fine structure of the cross peak multiplets provides information on the size of the J-couplings.

In the COSY spectrum coupling patterns are found easily; by starting from a given cross peak one searches for other cross peaks in the horizontal and vertical direction. A large number of two-dimensional spectra have the same general structure as the COSY spectrum with the difference that in general the meaning of the cross peaks varies depending on the type of experiment chosen. Important variants of the COSY experiment are the n-quantum filtered ($n = 2, 3 \dots$) COSY spectra that lead to a simplification of the spectrum. However, this simplification of the spectrum is paid for by a certain loss of information. In most cases two- or three-quantum filtered spectra are used because the signal-to-noise ratio decreases with high order filtering. A very important experiment is the relayed COSY experiment (abbreviation RELAY, synonymously used for RCT, relayed coherence transfer) where as well as the normal COSY peaks, signals occur that belong to nuclear spins whose coupling to each other is too small for detection but which both have a substantial J-coupling to the third nuclear spin. The complete network of J-coupled spins can be obtained with the total correlation spectroscopy (TOCSY), a method that is instrumentally more demanding than the RELAY experiment, but is more and more used because of its higher information content.

Essentially the same information as with the COSY experiment is obtained with the SECSY method (spin echo correlated spectroscopy) in which data acquisition is delayed by t_1 (pulse sequence $90°$-t_1-$90°$-t_1-acquisition) compared to the COSY spectrum. However, the geometry of the spectrum is different. In this case the central line corresponds to the one-dimensional spectrum. Corresponding cross peaks are situated on the straight line of the spectrum tilted by an angle of $135°$. Since it is not possible to obtain SECSY spectra with a pure absorption phase of the cross peaks, SECSY spectroscopy is actually only applied if storage capacity on the disk is not sufficient for a COSY experiment. A more sensitive detection can then be successfully done with a variant of the SECSY experiment, the super-SECSY.

The J-resolved 2D-spectroscopy is the 2D-experiment that was first performed in practice (pulse sequence: $90°$-t_1-$180°$-t_1-data acquisition). The J-resolved spectrum corresponds in its information contents to a 1D-spectrum in which the multiplets are turned perpendicular out of the ω_2 axis. Hence, the ω_2 axis contains the chemical shift of the centers of the multiplets, while the ω_1 direction represents the multiplet splitting. Since this spectrum does not provide information on which nuclear spins are coupled to each other, but only which spectra lines belong to one multiplet, J-resolved spectroscopy is now largely of historical interest.

An alternative to COSY spectroscopy is multiple-quantum spectroscopy. It has the advantage that it furnishes more information than COSY spectroscopy, but evaluating the spectra is more difficult. The two-quantum spectrum contains in addition to the connectivities between the directly coupled spins, the cross peaks of which are situated symmetrically to the two-quantum diagonal, peaks that occur without their symmetric partner. They provide information on nuclei that are coupled to another nuclear spin only indirectly, information analogous to that obtained by RCT spectroscopy.

2.3.3 Measurement of the Nuclear Overhauser Effect and Chemical Exchange

The nuclear Overhauser effect can be used to determine the spatial separation between two nuclear spins. The corresponding 2D-experiment is termed NOESY (nuclear Overhauser enhancement spectroscopy) and can also serve, as in the one-dimensional case, to investigate chemical exchange. The corresponding pulse sequence is shown in Fig. 2.14. NOESY spectra are similar to COSY spectra with the difference that here the cross peaks mean mutual NOE or chemical exchange. The intensity of these cross peaks depends on the chosen mixing time τ_m. Depending on τ_m the intensity of the cross peaks (the volume of the cross peaks) increases continuously, reaches a maximum value and decreases again to zero. The course of this process corresponds to the one found for the transient NOE in the one-dimensional case. The intensity I_{AB} of the cross peaks between the resonances of nucleus A and B are for short mixing times τ_m

$$I_{AB} \sim r_{AB}^{-6}\tau_m \quad . \tag{2.36}$$

In this equation r_{AB} denotes the distance of the nuclei between which the NOE takes place. In Chap. 3 we shall examine in more detail the dependence of the NOEs on the distance.

Two modifications of the NOESY experiments play a practical role. (1) The NOESY experiment with diagonal suppression where the intensity of the intense signal on the diagonal is reduced and (2) the ROESY experiment (rotating frame Overhauser enhancement spectroscopy) where the NOE is measured in the rotating coordinate system. This experiment is occasionally also termed CAMEL-SPIN. Experiment (1) permits a better evaluation of the spectra and experiment (2) plays an important role if the correlation times τ_c are in a range where the NOE in the laboratory system is close to one and therefore practically no Overhauser effect is detectable.

2.3.4 Correlation of Heteronuclear Resonances

Heteronuclear correlation, that is, the correlation of NMR spectra of different nuclear spins, has a number of attractive properties. In the first place, as usual with 2D-spectroscopy, it leads to a better resolution of superposed resonances. The correlation of two different NMR spectra, as for example the correlation of a

^{1}H spectrum (I spins) with the corresponding ^{13}C spectrum (S spins), facilitates assigning the resonance lines since in this case one has for the proton spectrum the additional information on the ^{13}C shift and vice versa. The assignment becomes trivial if one of the spectra is already completely interpreted. Furthermore, the two-dimensional correlation spectrum can be used for detecting, with higher sensitivity, signals from nuclei with lower magnetogyric ratios which therefore have a low NMR detectability. The simplest heteronuclear correlated experiment is set up very similarly to the COSY experiment. It consists again of two 90° pulses separated by the time t_1 in the frequency range of the I spins and is followed by recording the FID during the time t_2. The only difference is an additional 90° pulse in the frequency range of the S spins at the same time as the second 90° pulse of the I spins. The signal-to-noise ratio that can be obtained by different methods is given by

$$\mathrm{S/N} \sim \gamma_{\mathrm{exc}} \gamma_{\mathrm{obs}}^{3/2} (1 - e^{T_{\mathrm{R}}/T_{\mathrm{1exc}}}) \quad . \tag{2.37}$$

Here γ_{exc} is the magnetogyric ratio of the nuclear spins excited at first and γ_{obs} the magnetogyric ratio of the directly observed nucleus. In a repetitive experiment the signal-to-noise ratio depends also on the longitudinal relaxation time T_{1exc} of the excited spins and on the repetition time T_{R}. Hence, the most favorable signal-to-noise ratio can be expected in an experiment in which the coherence of the nuclear spin with the higher γ is first transferred to the one with the lower γ and is then transferred back for detection. According to (2.37) the ^{15}N resonances in a ^{15}N-^{1}H system can theoretically be detected with a 300-fold greater sensitivity, which is certainly a considerable gain.

3. NMR Spectroscopy of Biological Macromolecules

In this chapter we shall deal with NMR spectroscopy of biological macro-molecules, which is at present the most important application of biological NMR besides in vivo spectroscopy. The main problem in this case is frequently the assignment of the resonance lines, because in general with an increasing number of atoms that form a molecule the complexity of the NMR spectrum also increases. Furthermore, the linewidth of the resonance absorption lines increases with increasing molecular weight and the spectral resolution and the sensitivity are therefore markedly reduced.

The questions that are asked and their answers depend to some extent on the substance under investigation. We have therefore divided this chapter into sections according to the most important classes of biological macromolecules: the proteins, the nucleic acids, the polysaccharides and the lipids, each of which will be discussed in turn.

3.1 NMR Spectroscopy of Proteins

Polypeptides and proteins have many different functions in biological systems. They serve as structure components, as active elements in motion and transport, as storage reservoirs for metabolic energy, as biocatalysts and as transmitters of signals. In spite of these many different functions they all are built on the same structural principles. Like most biological molecules they are built of building blocks, the choice and sequence of which determines their specific properties. The basic structure elements of the proteins are the amino acids that are assembled into one or several linear chains.

The NMR of proteins is well developed and provides information on dynamical processes in proteins, and on their interaction with substrates and with macromolecular structures, that can be obtained by other methods only with difficulty or not at all. The most recent development in this field is the determination of the complete spatial structure of proteins using multi-dimensional NMR spectroscopy, in which the pioneering work was carried out by *Richard Ernst* and *Kurt Wüthrich*. The aims of this method appeared utopian only a few years ago; the importance and the extent of its impact on biochemistry and biophysics cannot yet be assessed.

3.1.1 Composition and Structure of Proteins

Of the many possible amino acids, proteins contain mainly the twenty L-amino acids, whose sequence in proteins is directly programmed in the genetic code (Fig. 3.1). The names of the amino acids are frequently abbreviated in a three- or a one-letter code (Table 3.1). The amino acids in proteins are linked by the peptide bond (Fig. 3.2), which is formed by joining the carboxyl group of one amino acid to the amino group of the next one in the chain.

The peptide bond is planar, with the participating atoms all situated in a plane. It occurs in two forms, the cis- and the trans-form. Most peptide bonds in proteins occur in the trans-configuration, the only exception being the peptide bond in which the imino group of the amino acid proline participates. In sim-

Table 3.1. Notations and properties of the standard amino acids

Name	Three-letter code	One-letter code	Molecular weight[a]	pK values of code side chains[b]
alanine	Ala	A	71	–
arginine	Arg	R	156	12.5
asparagine	Asn	N	114	–
aspartic acid	Asp	D	115	3.9
cysteine	Cys	C	103	8.3
glutamine	Gln	Q	128	–
glutamic acid	Glu	E	129	4.3
glycine	Gly	G	57	–
histidine	His	H	137	7.0 (6.2)
isoleucine	Ile	I	113	–
leucine	Leu	L	113	–
lysine	Lys	K	128	11.1
methionine	Met	M	131	–
phenylanaline	Phe	F	147	–
proline	Pro	P	97	–
serine	Ser	S	87	–
threonine	Thr	T	101	–
tryptophan	Trp	W	186	–
tyrosine	Tyr	Y	163	10.3
valine	Val	V	99	–

[a] Molecular weights of the subunits in the peptide chain at neutral pH.

[b] The pK values of cysteine and arginine were determined in free amino acids (Lehninger, 1983). The remaining pK values of the amino acids X were determined in the model peptides Gly-Gly-X-Ala by NMR (Bundi and Wüthrich, 1979). pK values in brackets: most probable values in proteins (Gross and Kalbitzer, 1988)

Fig. 3.1. The structures of the 20 standard amino acids. The amino acid units depicted correspond to the forms most frequently found in proteins at neutral pH. C atoms and H atoms are shown as white spheres, N, O and S atoms as shaded spheres

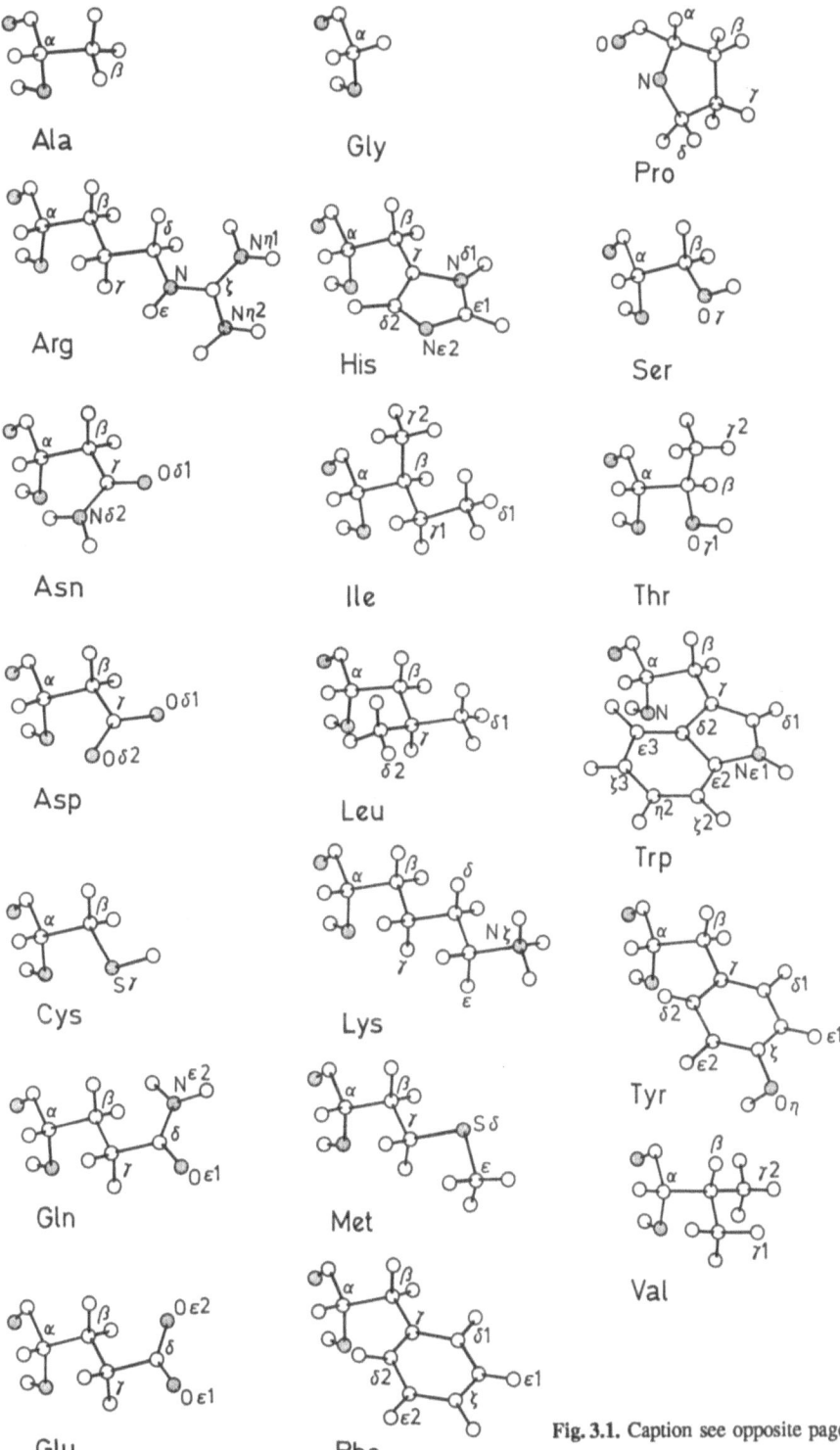

Ala

Gly

Pro

Arg

His

Ser

Asn

Ile

Thr

Asp

Leu

Trp

Cys

Lys

Tyr

Gln

Met

Val

Glu

Phe

Fig. 3.1. Caption see opposite page

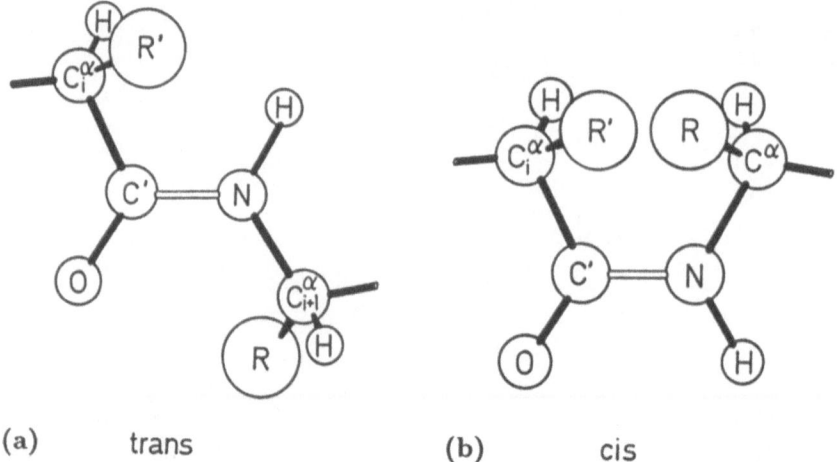

Fig. 3.2a,b. The peptide bond in trans configuration (a) and cis configuration (b). The C'-N bond has a partial double-bond character

ple peptides that contain proline, both isomers occur in solution in comparable concentrations. The equilibrium constant between the trans- and the cis-isomers is mostly in the range between 5 and 20. Assuming that the bond lengths and bond angles in proteins correspond to a good approximation to their standard values, the spatial structure of the protein depends only on the extent of the rotation around the single bonds. The corresponding angles are termed dihedral angles, and depending on their position in the amino acid, they are labelled with Greek letters. The three dihedral angles of the main chain are designated with the symbols φ, ψ, ω, and those of the side chain with χ^i (Fig. 3.3) where the superscript i describes the position in the side chain.

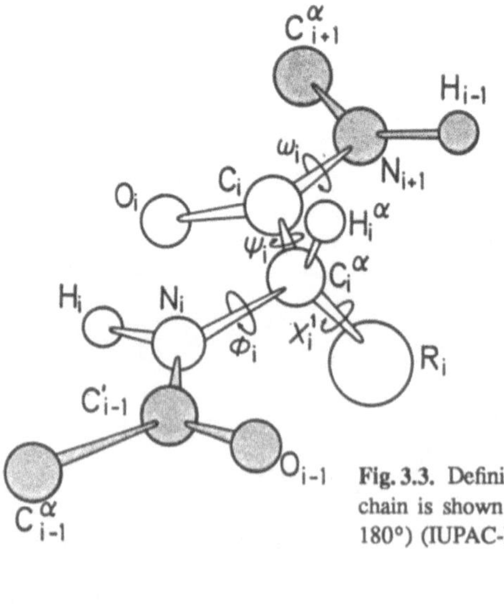

Fig. 3.3. Definition of the dihedral angles. The polypeptide chain is shown in its extended conformation ($\varphi = \psi = \omega = 180°$) (IUPAC-IUB, 1970)

The covalent structure of the amino acid side chain, that is, the sequence including possible modifications of the amino acids and disulphide bridges, is known as the primary structure of the protein. The secondary structure is then given by the local spatial arrangement of the atoms in the amino acid main chain, as determined by the dihedral angles. The tertiary structure is defined by the spatial folding of the total polypeptide chain. If a protein consists of several different polypeptide chains, the spatial configuration of the subunits with respect to each other is termed quarternary structure.

A number of secondary structures are found in proteins. They are characterized by the formation of particular patterns of hydrogen bonds between the amide

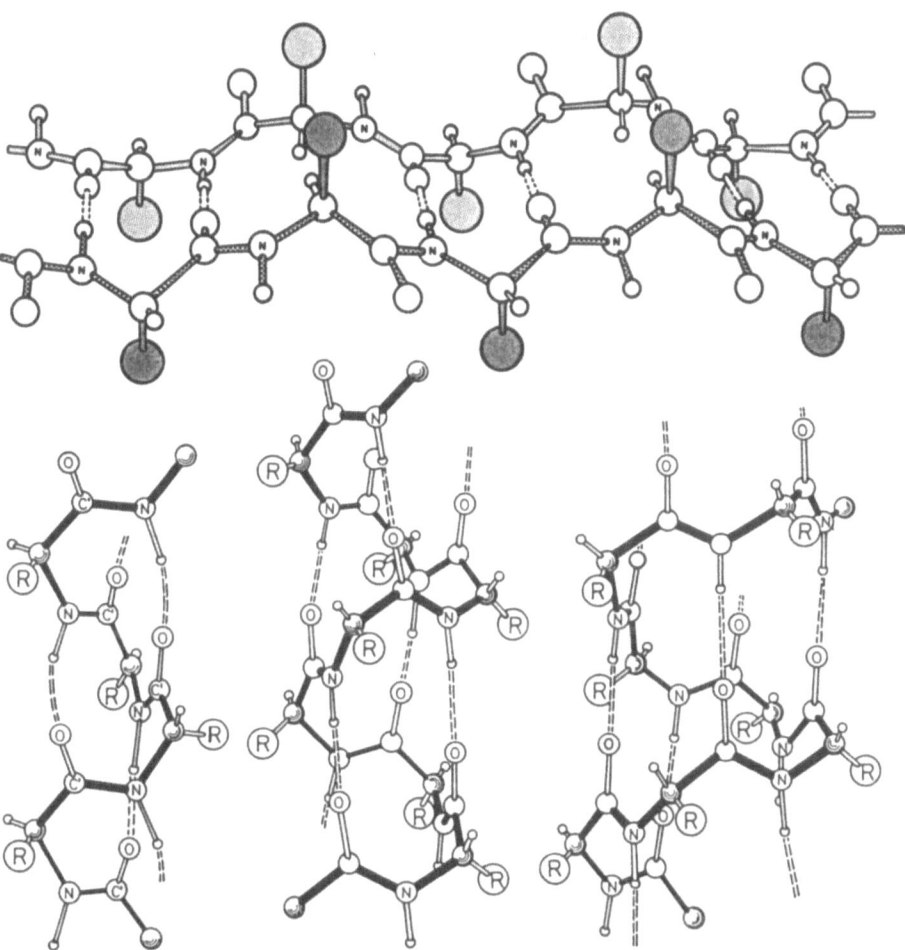

Fig. 3.4. Important secondary structure elements. The antiparallel β-pleated sheet (*top*), the 3_{10}-helix (*bottom left*), the right-handed α-helix (*bottom middle*) and the π-helix (*bottom right*). The hydrogen bonds are represented by broken lines, the side chains are symbolized by super atoms (R) (according to Schulz and Schirmer, 1979)

groups and the carbonyl groups of the main chain. Alternatively they may also be defined by the typical values of the dihedral angles φ and ψ. The two most important configurations are the right-handed α-helix and the β-pleated sheet (Fig. 3.4) that are found in most proteins. Other secondary structure elements that are also characterized by repeating (φ, ψ)-angles are the left-handed α-helix, the π-helix, the 3_{10}-helix and the collagen helix (Table 3.2). Not all combinations of the dihedral angles in the peptide chain occur with equal probabilities because of the steric hindrance of atoms, and the corresponding energies can differ markedly. Combinations of (φ, ψ)-angles with low potential energy are termed "allowed conformations", those with high energies are "forbidden conformations". The existence of such regions can be easily understood if the atoms in the simplest model are represented by hard spheres; for steric reasons some combinations of dihedral angles are simply impossible. These combinations can be represented clearly in the Ramanchandran plot in which an axis is assigned to each dihedral angle and allowed and forbidden regions are marked.

Table 3.2. Important secondary structure elements in proteins[a]

Secondary structure element	φ	ψ	hydrogen bonding pattern
parallel β-pleated sheet	-119	$+113$	CO_{i-1}-NH_j and NH_{i+1}-CO_j or
			CO_{j-1}-NH_i and NH_{j+1}-CO_i
antiparallel β-pleated sheet	-139	$+135$	CO_i-NH_j and NH_i-CO_j or
			CO_{i-1}-NH_{j+1} and NH_{i+1}-CO_{j-1}
3_{10}-helix	-76	-5	CO_i-NH_{i+3}
right-handed α-helix	-57	-47	CO_i-NH_{i+4}
left-handed α-helix	$+57$	$+47$	CO_i-NH_{i+4}
π-helix	-57	-70	CO_i-NH_{i+5}
collagen helix	-51	$+153$	CO_i-NH_{j-1} and NH_{i-1}-CO_k
	-76	$+127$	
	-45	$+148$	

[a] According to IUPAC-IUB (1970) and Ghélis and Yon (1982)

3.1.2 Dynamical Processes in Proteins

Biological polymers in solution do not have a rigid structure, but change their shape continuously. A protein has many possibilities of motion because of the large number of single bonds allowing free rotation. However, most proteins in their native form adopt comparatively densely packed, well-defined structures. Consequently, the number of possible motions is reduced since most combinations of dihedral angles are no longer possible without abandoning the native conformation, that is, without denaturation. Nevertheless, even in the native form, rotations of the side chains or groups within the side chains, in particular rotations of the methyl groups, collective torsional motions of the main chain as well as transitions between different native conformations are still possible. In addition, chemical exchange processes may occur: protons of charged groups, of amide, hydroxyl and sulphydryl groups may dissociate from and rebind to the protein, as may also substrates and products of enzymes. All these processes are reflected in the NMR spectrum of the protein.

The planar ring systems of the amino acids histidine, tyrosine, phenylalanine and tryptophan can in principle rotate around the C^α-C^β- and C^β-C^γ-axes. However, it is easy to imagine that this rotation is frequently hindered by the interaction with other atoms and becomes completely impossible in the core of the protein. This process can be observed with ^{1}H-NMR. The electronic clouds of the ring systems of neighboring atoms shield the external magnetic field in a different manner, an effect that can be observed not only in the nucleus within the electronic cloud, but also with the neighboring atoms. This effect depends, of course, on the spatial position of the protons observed with respect to the rest of the molecule. The rings of the amino acids tyrosine and phenylalanine are symmetric with respect to the twofold axis around this C^β-C^γ-bond (Fig. 3.1), that is, in symmetrical surroundings the chemical shifts of the two δ protons and the two ε protons are identical and hence only one resonance line is found for both pairs of protons. Only an asymmetry of the surroundings that is in general expected for proteins may lead to different chemical shifts for the symmetrically arranged protons and thus to a splitting of the resonance lines. However, if the ring system rotates fast around its axis, these effects are averaged out and only one single averaged resonance line appears (fast exchange). Since the speed of the rotation depends on the temperature, the transition from slow to fast exchange with increasing temperature can sometimes be directly observed in the NMR spectrum.

A good example is given in Fig. 3.5 that shows the dynamics of the phenylalanine and tyrosine rings in basic pancreatic trypsin inhibitor (BPTI): at high temperatures the δ-(2,6-) and the ε-(3,5-) protons are equivalent in all tyrosine and phenylalanine rings. If the temperature is reduced, two of the aromatic residues, Phe-45 and Tyr-35, change their spectrum because the speed of the rotation is no longer sufficient for averaging the anisotropy. A simulation of these spectra provides the rates of rotation and the corresponding activation energies. Both rings are situated in the densely packed inner part of the protein; it is surprising that they can nevertheless rotate rather fast. However, there is much indication that they do not rotate steadily but rather flip fast by 180°. The remaining aromatic ring systems in BPTI do not show signs of a hindrance of their rotation in the whole temperature range; their spectra remain typical for fast exchange, a property that one is surprised to find with most proteins.

Proteins are frequently found in several different conformations; the extreme conformation or class of conformations is the completely denatured state, the random coil in which by definition order no longer exists. The transition from the native conformation into this state can easily be followed by ^{1}H-NMR. In the folded state the ^{1}H-NMR spectrum of proteins is strongly structured and

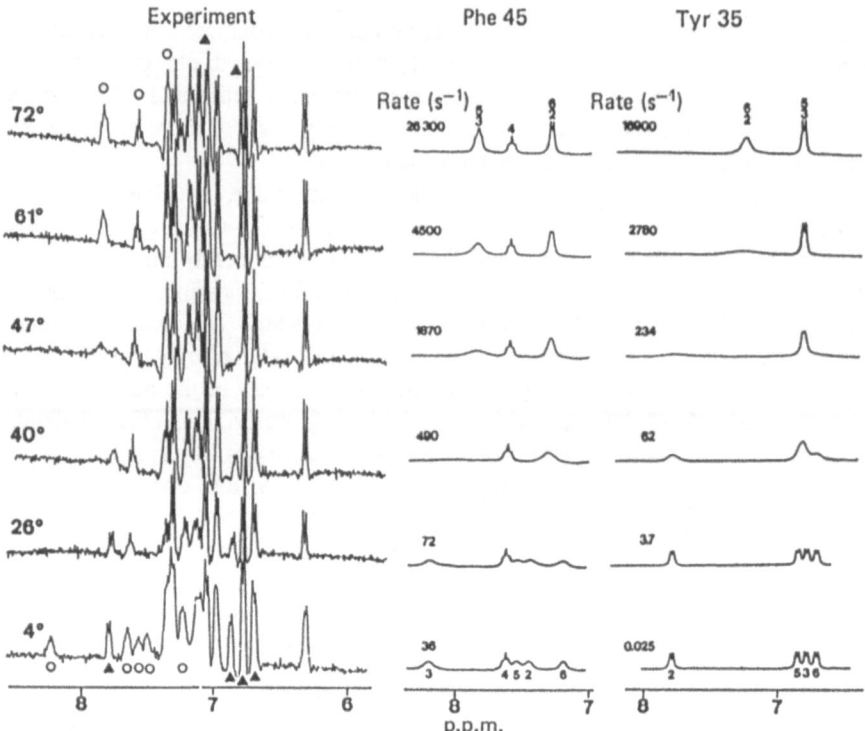

Fig. 3.5. Reorientation of aromatic rings in proteins. The rotational motions of the aromatic rings of Phe-45 and Tyr-35 are hindered in the bovine pancreatic trypsin inhibitor (BPTI). Increasing the temperature results in a transition from slow to fast exchange. The resonances of Phe-45 are labelled with (o) in the experimental spectrum, those of Tyr-35 with (▲). On the right side the simulated spectra are shown together with the corresponding rotation rates (Wüthrich and Wagner, 1978, with permission)

it is not possible to predict its shape precisely even if the spatial structure is known. In contrast to this, the spectrum in its denatured state is simply given to first approximation by the amino acid composition. The chemical shifts of the individual components then agree very well with those that are found in model substances. A model system that is generally accepted consists of tetrapeptides Gly-Gly-X-Ala in which X is just the amino acid in question (Table 3.3). However, these values are still slightly modified even in the random coil by the influences of the nearest neighbors in the sequence, but by definition not by influences from the secondary and tertiary structure. Even with denatured proteins the NMR spectrum often still shows deviations from the typical random coil values. This indicates, of course, that in these states some local order can still exist. Conversely, the NMR spectrum is a good indicator of whether or not a protein is really completely denatured.

Figure 3.6 shows the ^{1}H-NMR spectrum of factor III[lac], a protein that participates in the vectorial phosphorylation of the sugar lactose. In this process lactose

Table 3.3. Chemical shifts in ^{1}H-NMR spectra of a random-coil model peptide[a]

Amino acid residue X	H	$H\alpha$	$H\beta$	$H\gamma$	other atoms				
Ala	8.25	4.35	1.39						
Arg	8.27	4.38	1.79	1.70	$H\delta$	3.32	$H\varepsilon$	7.17	$H\eta$ 6.62
			1.89	1.70	$H\delta$	3.32			
Asn	8.75	4.75	2.75		$H\delta2$	6.91			
			2.83		$H\delta2$	7.59			
Asp	8.41	4.76	2.75						
			2.84						
Cys	8.31	4.69	2.96						
			3.28						
Gln	8.41	4.37	2.01	2.38	$H\varepsilon2$	6.87			
			2.13	2.38	$H\varepsilon2$	7.59			
Glu	8.37	4.29	1.97	2.28					
			2.09	2.31					
Gly	8.39	3.97							
		3.97							
His	8.41	4.63	3.20		$H\delta2$	7.14	$H\varepsilon1$	8.12	
			3.26						
Ile	8.19	4.23	1.90	1.19	$H\gamma2$	0.95	$H\delta1$	0.89	
				1.48					
Leu	8.42	4.38	1.65	1.64	$H\delta$	0.90			
			1.65		$H\delta$	0.94			
Lys	8.41	4.36	1.76	1.45	$H\delta$	1.70	$H\varepsilon$	3.02	$H\zeta$ 7.52
			1.85	1.45	$H\delta$	1.70	$H\varepsilon$	3.02	
Met	8.42	4.52	2.01	2.64	$H\varepsilon$	2.13			
			2.15	2.64					
Phe	8.32	4.66	2.99		$H\delta1$	7.30	$H\varepsilon1$	7.39	$H\zeta$ 7.34
			3.22		$H\delta2$	7.30	$H\varepsilon2$	7.39	
Pro[b]		4.44	2.02	2.03	$H\delta$	3.65			
			2.28	2.03	$H\delta$	3.68			
Ser	8.38	4.50	3.88						
			3.88						
Thr	8.24	4.35	4.22	1.23					
Trp	8.09	4.70	3.19		$H\delta1$	7.24	$H\varepsilon1$	10.22	$H\zeta2$ 7.50
			3.32		$H\eta2$	7.24	$H\varepsilon3$	7.65	$H\zeta3$ 7.17
Tyr	8.18	4.60	2.92		$H\delta1$	7.15	$H\varepsilon1$	6.86	
			3.13		$H\delta2$	7.15	$H\varepsilon2$	6.86	
Val	8.44	4.18	2.13	0.94					
				0.97					

[a] Data from the tetrapeptide Gly-Gly-X-Ala in water at 308 K and pH 7.0 (Wüthrich, 1986)

[b] Proline in trans configuration

is transported through the cell membrane of the bacterium *Staphylococcus aureus* and simultaneously phosphorylated. Four proteins participate in this process, enzyme I, HPr, factor III and enzyme II. In a cascade the phosphate group that originates from phospho*enol*pyruvate is transferred from enzyme I via HPr, factor

III and enzyme II to the lactose, which in this process is simultaneously passed through the cell membrane. Because of its function this membrane transport system is also termed phospho*enol*pyruvate dependent phosphotransferase system (PTS). Factor III^{lac} is a protein of medium size and consists of three identical subunits, each of which is composed of 103 amino acids. The NMR spectrum is typical for a protein of this size in the native state. Most details in the spectrum are not resolved because a large number of resonance lines are superposed. Only in the outer parts of the spectrum, the low field part above 6 ppm and the high field part below 0.7 ppm, the signals of individual groups can be resolved. The resonances marked with d_1, d_2 and d_3 belong to high field shifted methyl resonances; the resonances in the low field range correspond to the ring protons of the 4 histidine, 5 phenylalanine and 2 tyrosine residues (Fig. 3.6). Figure 3.7

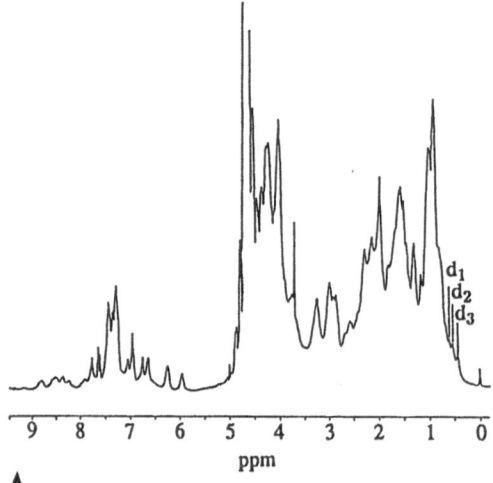

Fig. 3.6. 360 MHz ^{1}H-NMR spectrum of the native factor III^{lac} from *Staphylococcus aureus* in D$_2$O at 308 K (Kalbitzer et al., 1981, with permission)

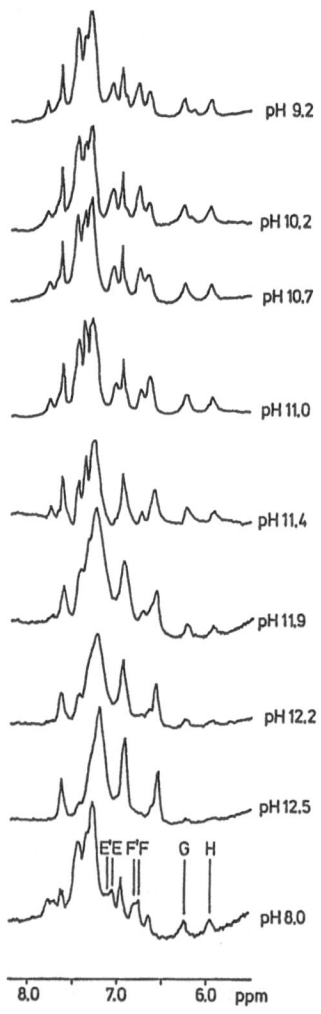

Fig. 3.7. Denaturation of factor III^{lac}. This factor denatures ▶ with increasing pH. The bottom spectrum was the last one measured in the series. E, E', F, F' label tyrosine resonances in the native and partially denatured state (Kalbitzer et al., 1981, with permission)

shows the changes in the low field range of the NMR spectrum due to alkaline denaturation. The denaturation is shown in particular by the two resonance lines that are labelled with G and H in the lowest spectrum. Starting from pH 11.4 their intensity decreases until they disappear completely at pH 12.5. The resonance lines of the ring protons are situated at their random coil positions. The peak at 7.7 ppm corresponds to the $\varepsilon 1$-(2-) protons of the histidines, the large peak at approximately 7.3 ppm to the ring protons of the phenylalanines, that at 7.0 ppm to the $\delta 2$-(4-) protons of the histidines and to the δ-(2,6-) protons of the tyrosines, and the one at 6.6 ppm to the ε-(3.5-) protons of the tyrosines. If the pH value is reset to its physiological value after the alkaline denaturation, the protein is renatured. However, a comparison of the NMR spectra shows that the renaturation is not complete.

Changes of conformation after binding to the substrate are part of many theories on enzymatic catalysis and its regulation. A good example of this is adenylate kinase, which catalyzes the reaction

$$Mg.ADP + ADP \rightleftharpoons Mg.ATP + AMP \tag{3.1}$$

and assures under physiological conditions that with high ATP consumption the ADP level remains low and simultaneously additional ATP is made available for the hydrolysis. For adenylate kinase two very different structures are found by X-ray structure analysis, probably corresponding to the open form before and the closed form after the substrate binding. One would expect that different NMR spectra would result from such very different conformations. Extreme changes of the ^{1}H-NMR spectrum of the human adenylate kinase are indeed found if the bisubstrate analogue Ap$_5$A binds (in Ap$_5$A two adenosine molecules are connected by a chain of five phosphate groups) (Fig. 3.8).

Since Ap$_5$A binds very strongly to the enzyme and is only very slowly exchanged ($k_{off} < 70\,\mathrm{s}^{-1}$), it is possible to observe the resonance lines of various amino acid residues both in free and in complexed adenylate kinase. For instance, at half saturation of the active center the resonance line of histidine 36 is split into two lines of equal intensity which correspond to the histidine residue in the substrate-free and in the complexed enzyme. However, not all spectral changes are due to the structural isomerization of the adenylate kinase: the substrate itself can also be directly responsible for a part of these changes. Both factors cannot easily be separated, in particular, if the substrate contains aromatic rings. As we shall see below, the corresponding ring current shift also causes significant changes of the resonance frequencies for nuclei that are comparatively distant from the ring center.

Ionizable groups frequently play an important role in catalytic processes. A fundamental requirement for understanding the mode of action is to know the state of charge of these groups under physiological conditions. At a given pH value of the surroundings the charge of these groups is determined by their pK values. It is this pK value, the equilibrium between protonation and deprotonation, that can be elegantly determined by NMR for individual groups in

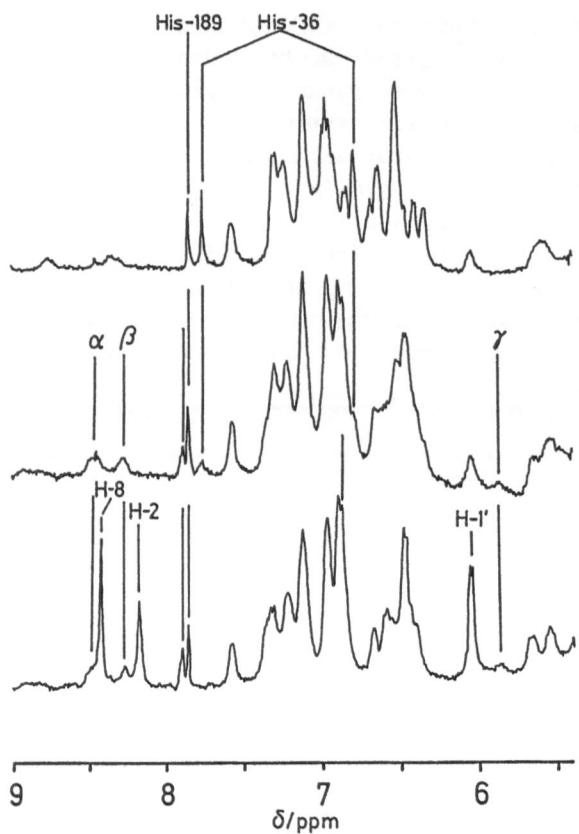

Fig. 3.8. The substrate induced conformational change of adenylate kinase. 360 MHz ^{1}H-NMR spectrum of human adenylate kinase in the absence of substrate (*top spectrum*), after addition of Ap$_5$A at the ratio 1 : 2 (*middle spectrum*) and after addition of Ap$_5$A in excess (Kalbitzer et al., 1982a, with permission). The resonances of the ring protons of His-36 and His-189 are labelled. α, β and γ label the resonances of the H-1', H-2 and H-8 protons of bound Ap$_5$A

the protein if the corresponding signals in the spectrum can be identified. This is particularly simple in the spectral range of the resonance frequencies of the ring protons. In this region there are only comparatively few superpositions of resonance lines because most proteins contain only a small number of aromatic amino acids. An amino acid that participates in many catalytic processes and has its resonance frequency in this range is histidine. Its signals can frequently be resolved even in large proteins.

In the protein components of the phospho*enol*pyruvate dependent phosphotransferase system described above, the phosphonate group which they transfer is transiently bound to a histidine residue in the active center. The histidine signals could be identified in factor III and in the HPr protein. For the histidine in the active center a pK value of about 6 for the two proteins was found with NMR, that is, the histidine ring is deprotonated at physiological pH. To find out whether this is accidental, the HPr proteins of different microorganisms were investigated (Fig. 3.9). The NMR characteristics of these histidines are amazingly constant, although the amino acid compositions are markedly different. However, this similarity cannot be seen easily in Fig. 3.9 because the spectra were measured at different pH values. The pK value found with a pH titration is

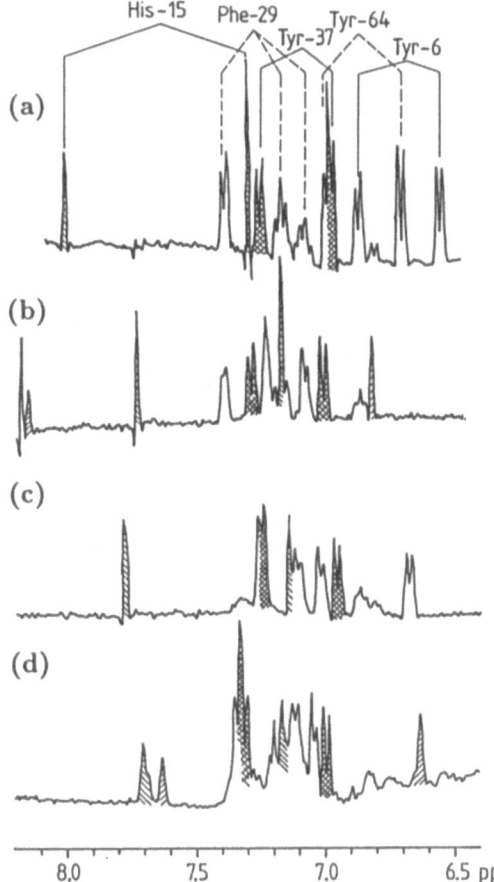

His-15   Phe-29   Tyr-64
              Tyr-37   Tyr-6

(a)

(b)

(c)

(d)

8.0   7.5   7.0   6.5 ppm

Fig. 3.9a–d. The histidine residue in the active centers of HPr proteins. 360 MHz ^{1}H-NMR spectra of HPr proteins of different grampositive microorganisms. Only the downfield part of the NMR spectrum with the resonances of the ring protons is shown. HPr from *Staphylococcus aureus* (a), from *Streptococcus lactis* (b), from *Bacillus subtilis* (c) and from *Streptococcus faecalis* (d). The active-center histidine, His-15, has very similar NMR parameters for all proteins shown. Temperature 308 K, pH values 6.2 (a), 7.7 (b), 8.4 (c) and 8.8 (d). The resolution of the resonance lines was enhanced by application of a Gaussian filter (Kalbitzer et al., 1982b, with permission). The positions of the aromatic residues given here correspond to the latest corrected amino acid sequence

in fact smaller than 6.1 for all HPr proteins investigated (Fig. 3.10). The same experiments can also be performed in the phosphorylated state. Here unusually high pK values between 7.8 and 8.6 are found for factor III and HPr protein, that is, the phosphohistidines are protonated at physiological pH. The results are schematically depicted in Fig. 3.11. Note that the phosphonate groups in the HPr protein and in factor III are bound to different nitrogen atoms in the ring. In ^{1}H-NMR the phosphorylation of a histidine in the protein leads to typical changes of the histidine resonance frequency. This not only permits the conclusion that a certain histidine is phosphorylated, but also makes it possible to decide in which position in the ring the phosphonate group is bound.

The interaction of proteins that is reflected in the ^{1}H-NMR spectrum can be investigated with comparatively simple methods. Here, the proteins of PTS are again a good example. When transferring the phosphonate group, a complex of HPr and factor III is formed as intermediary. Both proteins can in principle interact in the phosphorylated or unphosphorylated state. One expects that the

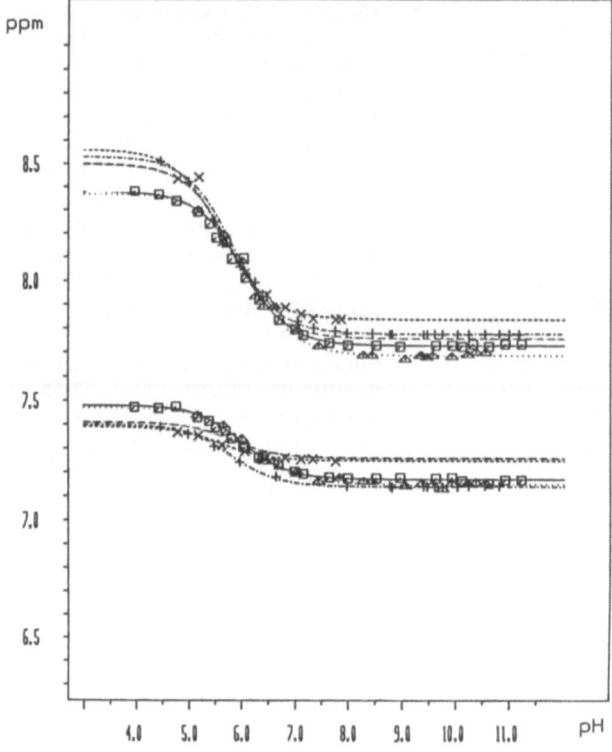

Fig. 3.10. pH dependence of chemical shifts of the ring protons of His-15 in HPr proteins from *S. lactis* (□), *B. subtilis* (+), *S. faecalis* (△), *E. coli* (×) and *S. aureus* (- - -). The curves shown were calculated with the modified Henderson-Hasselbalch equation (Kalbitzer et al., 1982b, with permission)

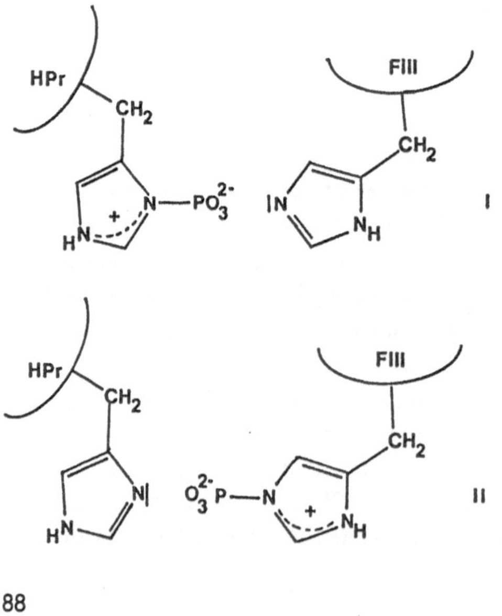

Fig. 3.11. The ionization state of the histidine rings in the active centers of HPr protein and factor III before (I) and after (II) transfer of the phosphonate group at pH 7.4 as deduced from NMR studies (Kalbitzer et al., 1981, with permission)

proteins recognize each other in each of these combinations, but that the interaction is weaker if both components are either both phosphorylated or both not phosphorylated. If NMR spectra of unphosphorylated factor III are recorded and the chemical shift of the resonance line of its histidine in the active center is observed, depending on the concentration of unphosphorylated HPr added, the histidine resonance line is continuously shifted to higher field. This is a typical behavior for the case of fast exchange between two surroundings, histidine in free factor III and in factor III complexed with HPr. From the condition for fast exchange $|\Delta\omega\tau_e| \ll 1$, relation (2.10), it is possible to estimate a lower limit for the exchange rate $1/\tau_e \gg 2300\,\mathrm{s}^{-1}$. Fitting the data to equation (2.12) modified so that the populations follow the law of mass action, it is even possible to estimate a lower limit for the association constant between the two proteins ($K > 5000\,M^{-1}$).

An interesting phenomenon is the exchange of the amide protons of the peptide backbone with the surrounding water. We have already become acquainted with this phenomenon as a tool for simplifying spectra when a protein is dissolved in D_2O in which the amide protons exchange with the deuterons of the solvent. The rate constant with which this exchange takes place depends strongly on structural features. If the exchange is very slow, it can be directly followed with time from the decrease of the line intensity. Higher exchange rates can be elegantly determined with the saturation transfer experiment in H_2O already discussed in Sect. 2.2.5. However, if the amide protons exchange so fast that they do not produce a resonance line separated from the signal of the water (fast exchange), no simple possibility exists to measure the exchange rates. Detailed investigations were made for the exchange of amide protons in BPTI in which the large majority of its resonance lines in the 2D-NMR spectrum could be assigned. The exchange rates that were found cover a very wide time range. The highest rates of about $2\,\mathrm{s}^{-1}$ (pH 3.5 and 36° C) correspond approximately to those that are also observed in unstructured peptides. The smallest rates are of the order of $1 \times 10^{-5}\mathrm{s}^{-1}$, which means that one must wait for days for this exchange to take place. The exchange itself depends on the one hand directly on the pH value of the solvent because it is catalyzed by H_3O^+ ions and OH^- ions (Fig. 3.12); on the other, it depends strongly on the spatial structure of the protein. Predominant factors in this connection are the water accessibility and electrostatic effects of neighboring groups. This is shown in Fig. 3.13 with the example of BPTI of which a very good X-ray structure exists in addition to the NMR assignments. Properties such as water accessibility and hydrogen bond formation can easily be deduced from these X-ray structure data, but one has to be somewhat cautious when transferring these data to the solute state. It can be seen that a certain correlation exists between water accessibility and the size of the amide exchange rates, that is, small exchange rates indicate a reduced accessibility of water protons to the peptide bond.

In principle, the relaxation times of the protons in the protein depend on motional processes in the protein and can therefore provide information on these

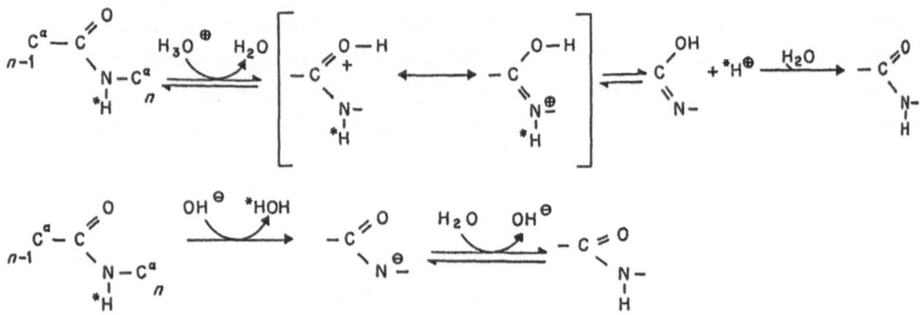

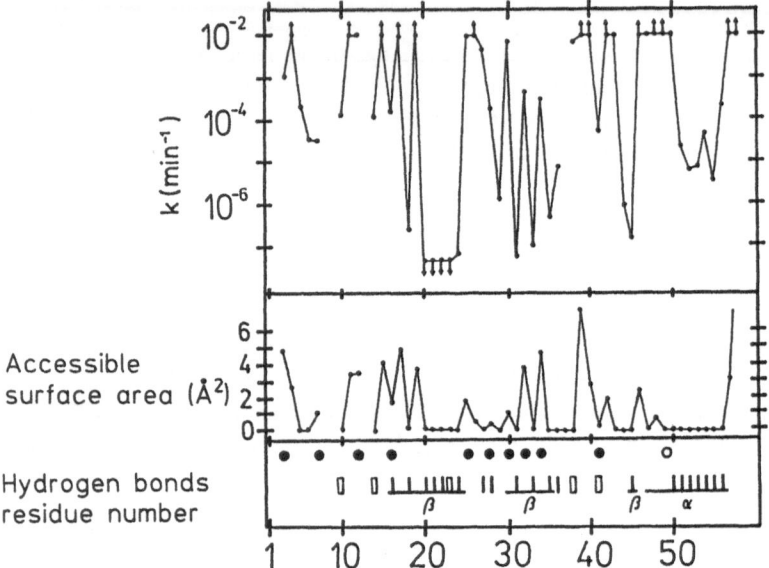

Fig. 3.12. Mechanism of the acid (*top*) and base (*bottom*) catalyzed exchange of amide protons of the peptide bond (Tüchsen and Woodward, 1985, with permission)

k (min⁻¹)

Accessible surface area (Å²)

Hydrogen bonds residue number

Fig. 3.13. Exchange of amide protons and structural properties. Relationship of the exchange rates k (measured at pH 3.6 and 36° C) of the amide protons of BPTI with the water accessibility of the peptide bonds and the existence of hydrogen bridges. Hydrogen bonds to the carbonyl oxygen of the main chain (I) or to atoms of the side chain or bound water molecules (☐). (α) α-helical region, (β) β-pleated sheet (Wagner, 1983, with permission)

processes (see Section 2.2.1). In the simple case of dipolar relaxation through rotational diffusion for two unequal nuclei A and B with spin $I = \frac{1}{2}$ the longitudinal relaxation rate is given by ϱ_A (equations 2.22, 2.27–2.31). Explicitly, neglecting external relaxation for spin A, one obtains

$$\frac{1}{T_1^{AB}} = \left(\frac{\mu_0}{4\pi}\right)^2 \frac{\hbar^2 \gamma_A^2 \gamma_B^2}{10 r_{AB}^6} \left(\frac{\tau_{\text{rot}}}{1 + (\omega_A - \omega_B)^2 \tau_{\text{rot}}^2} + \frac{3\tau_{\text{rot}}}{1 + \omega_A^2 \tau_{\text{rot}}^2}\right.$$
$$\left. + \frac{6\tau_{\text{rot}}}{1 + (\omega_A + \omega_B)^2 \tau_{\text{rot}}^2}\right) . \tag{3.2}$$

Analogously one obtains for the transversal relaxation rate

$$\frac{1}{T_2^{AB}} = \left(\frac{\mu_0}{4\pi}\right)^2 \frac{\hbar^2 \gamma_A^2 \gamma_B^2}{20 r_{AB}^6} \left(4\tau_{\text{rot}} + \frac{\tau_{\text{rot}}}{1+(\omega_A - \omega_B)^2 \tau_{\text{rot}}^2} + \frac{3\tau_{\text{rot}}}{1+\omega_A^2 \tau_{\text{rot}}^2}\right.$$

$$\left. + \frac{6\tau_{\text{rot}}}{1+\omega_B^2 \tau_{\text{rot}}^2} + \frac{6\tau_{\text{rot}}}{1+(\omega_A + \omega_B)^2 \tau_{\text{rot}}^2}\right) . \tag{3.3}$$

However, reaching equilibrium again after a perturbation also depends on the relaxation times of the nuclear spin B, so that in general the time dependence has to be described by a system of differential equations. For two equivalent nuclei with spin $\frac{1}{2}$ this results in a simple exponential relation with the relaxation rates

$$\frac{1}{T_1^{AA}} = \left(\frac{\mu_0}{4\pi}\right)^2 \frac{\hbar^2 \gamma_A^4}{10 r_{AA}^6} \left(\frac{3\tau_{\text{rot}}}{1+\omega_A^2 \tau_{\text{rot}}^2} + \frac{12\tau_{\text{rot}}}{1+4\omega_A^2 \tau_{\text{rot}}^2}\right) \tag{3.4}$$

and

$$\frac{1}{T_2^{AA}} = \left(\frac{\mu_0}{4\pi}\right)^2 \frac{\hbar^2 \gamma_A^4}{20 r_{AA}^6} \left(9\tau_{\text{rot}} + \frac{15\tau_{\text{rot}}}{1+\omega_A^2 \tau_{\text{rot}}^2} + \frac{6\tau_{\text{rot}}}{1+4\omega_A^2 \tau_{\text{rot}}^2}\right) . \tag{3.5}$$

For equivalent nuclear spins the longitudinal relaxation rate passes through a maximum with increasing $\omega_A \tau_{\text{rot}}$ and then decreases again. The transverse relaxation rate, on the other hand, increases continuously until finally the theory that is based on homogeneous lines is no longer applicable, since for these long correlation times the spectrum consists of the inhomogeneous lines of the solid state NMR. (Fig. 3.14).

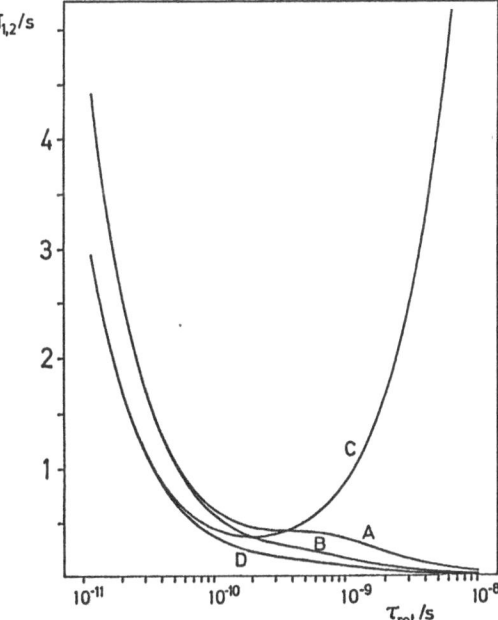

Fig. 3.14. Dependence of the longitudinal and transverse relaxation times on the rotational correlation time τ_{rot}. The relaxation times T_1^{AB} (A), T_2^{AB} (B), T_1^{AA} (C) and T_2^{AA} (D) are plotted as a function of τ_{rot}. For the calculation a resonance frequency of 500 MHz, a line separation of 1 ppm and a distance r_{AB} between the nuclear spins of 0.175 nm (corresponding to the distance of two methylene protons) has been assumed

The rotational correlation time of a spherical molecule can be calculated with the Stokes-Einstein relation

$$\tau_{\text{rot}} = \frac{4\pi\eta a^3}{3kT} \quad , \tag{3.6}$$

where η is the viscosity of the solvent, a the effective radius of the complex, k the Boltzmann constant and T the absolute temperature. A good measure for the volume V of a spherical molecule with the molecular weight M is

$$V = MV' = \frac{4\pi a^3}{3} \quad . \tag{3.7}$$

In proteins the partial volume V' is usually about $1.20 \times 10^{-30} \text{m}^3$.

Equations (3.6) and (3.7) can, of course, be only rough approximations for the rotational correlation time because normally real biological molecules are neither precisely spherical nor completely rigid. However, the values calculated with these equations provide a rather good estimation of the upper limit of the correlation time τ_{rot}.

Great hopes were initially placed in the measurements of the relaxation times, from which one hoped to elucidate in detail the motional processes in proteins. But, as could have been predicted, it turned out that the system possesses too many degrees of freedom and unknown factors to determine the complex motions in the protein independent of a model from the limited information available. This is particularly true for the ^{1}H-NMR where the relaxation times depend on the relative motion of a large number of protons and their mutual relaxation interaction. In principle it is easier to interpret the ^{13}C relaxation because it is essentially determined solely by the nuclear spins of the protons directly bound to it. However, here too the results depend strongly on the motional model chosen; at most, they can help in rejecting certain models as inadequate or in classifying them as consistent with the data.

3.1.3 Determination of Interatomic Distances

Measuring the distance between two atoms is very important for determining the spatial structure with NMR methods. The most important source for obtaining information on these distances is at present the measurement of the nuclear Overhauser effect. Additional information may in suitable cases be drawn from the interaction with paramagnetic centers and from ring current effects. In the following we shall discuss in greater detail how information on the distances can be derived from these interactions.

The basic features of the nuclear Overhauser effect were already discussed in Sect. 2.2.6. For proteins, one-dimensional methods to measure the nuclear Overhauser effect are only of limited value because very few regions in the ^{1}H-NMR spectrum are sufficiently well resolved that individual resonance lines can be selectively irradiated. Usually, a 2D experiment has to be performed, the NOESY experiment mentioned above, in which the superpositions do not play more than a minor role. For a two-spin system the time dependence of the nuclear Overhauser effect can be directly obtained by integration of the differential equations (2.20) and (2.21). The time dependence in the transient Overhauser experiment and in the NOESY experiment is given by

$$\text{NOE} = \frac{M_z^A(\tau_m)}{M_0^A} = \frac{I_z^A(\tau_m)}{I_0^A}$$

$$= -\frac{1}{2}\frac{\sigma_{AB}}{|\sigma_{AB}|}(1 - e^{-2|\sigma_{AB}|\tau_m})e^{-(\varrho_A - |\sigma_{AB}|)\tau_m} \quad . \tag{3.8}$$

Here again M_z^A and I_z^A are the z magnetization and the z component of the nuclear spin A, and τ_m is the time during which spin A and spin B have exchanged their z magnetization. M_0^A and I_0^A are the corresponding quantities in thermal equilibrium and σ_{AB} and ϱ_A are the spin-lattice and cross relaxation rates, respectively. Hence the NOE observed increases at first and then approaches zero for large times τ_m. Figure 3.15 shows as an example the time dependence measured for several protons in BPTI. TOE measurements exhibit a similar time dependence. However, for long times one does not obtain zero, but the equilib-

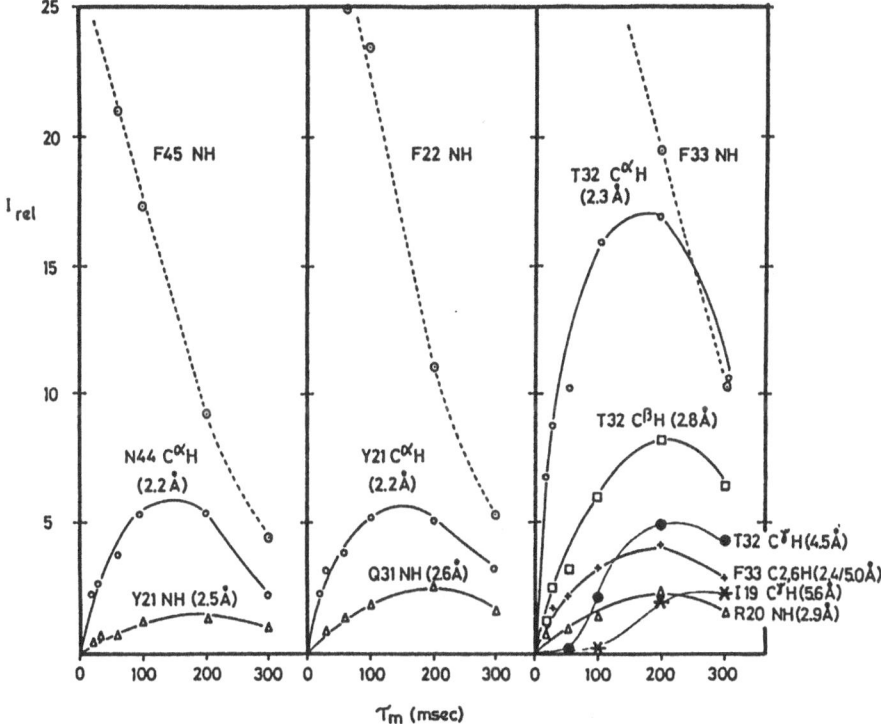

Fig. 3.15. Dependence of the cross peak amplitude on the mixing time τ_m in the NOESY spectrum of BPTI. The broken lines represent the amplitudes of the diagonal peaks of the amide protons of Phe-45, Phe-22 and Phe-33. The solid lines represent the NOEs between these nuclei and their neighbors in the 3D structure (Kumar et al., 1981, with permission)

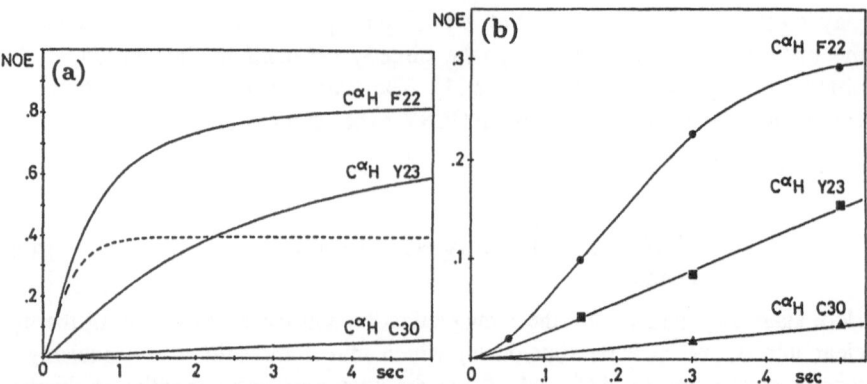

Fig. 3.16a,b. Dependence of NOEs on the duration of saturation in the TOE experiment. Saturation of the amide resonance of Tyr-23 of BPTI and observation of the resonances of the $H\alpha$ atoms of Phe-22, Tyr-23 and Cys-30. (a) Simulation of the time dependence, (b) expansion of (a) showing the experimental data (Dubs et al., 1979, with permission)

rium value of the NOE (see Fig. 3.16). Since in the protein, however, not merely two isolated spins but a large number of spins interact with each other, the equations developed for a two-spin system must be generalized. The generalization of (2.20) and (2.21) to N spins is obvious. Observing spin A gives

$$\frac{dM_z^A}{dt} = -\varrho_A(M_z^A - M_0^A) - \sum_{\substack{j=1 \\ j \neq A}}^{N} \sigma_{Aj}(M_z^j - M_0^j) \quad . \tag{3.9}$$

Since the size of M_z^j is influenced by the entire system, the course of the magnetization of nuclear spin A depends in principle on the entire system. Of course, the ultimate aim of the measurement is the determination of the interatomic distances r_{Aj} between atoms A and j. It can be shown that the change of the NOEs is again proportional to r_{Aj}^{-6} for very short mixing times τ_m ($\tau_m \to 0$). For the cross peak (more precisely, the volume of the cross peak) in the NOESY spectrum and for the NOE from one-dimensional methods one obtains

$$\frac{dNOE_{Aj}}{d\tau_m}(\tau_m \to 0) = -\sigma_{Aj} \quad . \tag{3.10}$$

For long correlation times σ_{rot}^{Aj} and predominating dipolar relaxation $\sigma_{Aj} = W_2^{Aj} - W_0^{Aj}$ becomes

$$\sigma_{Aj} = -\left(\frac{\mu_0}{4\pi}\right)^2 \frac{\hbar^2}{10} \gamma_A^2 \gamma_j^2 \tau_{rot}^{Aj} \langle r_{Aj}^{-6} \rangle \quad . \tag{3.11}$$

Here $\langle r_{Aj}^{-6} \rangle$ characterizes the average value over the interatomic distances. In general, these distances vary at a given time from molecule to molecule if the molecules are not completely rigid.

The correlation times are usually not so long that (3.11) can be applied, therefore the dependence of σ_{Aj} on the correlation time must also to be taken into account. In addition, equations (2.27–2.32) are strictly valid only for simple rotational diffusion. Since normally neither the precise correlation times nor the exact correlation functions are known, one can only try to find suitable approximations. One possibility for doing this is to use for calibration the NOE between two nuclear spins C and D in the macromolecule, the distance between which is known from the covalent structure. Assuming that for spin A and j the same correlation functions and correlation times are valid as for spin C and D, one obtains for the distance

$$\langle r_{Aj}^{-6} \rangle = r_{CD}^{-6} \frac{\mathrm{NOE}_{Aj}}{\mathrm{NOE}_{CD}} \quad . \tag{3.12}$$

However, two difficulties for determining the distance remain: (1) The assumption on the correlation functions is certainly valid only approximately at best, and (2), even if (1) is sufficiently satisfied, it renders only the average value over r_{Aj}^{-6}, but not the really interesting average distance r_{Aj} that may differ markedly from the one calculated from $\langle r_{Aj}^{-6} \rangle$.

As noted above, determining the distances from the NOEs appears to be very inaccurate. However, for most practical applications this is not as serious as it might seem. When measured for small mixing times the NOE always decreases very rapidly with increasing distances (proportional to the sixth power!). Hence NOEs can usually only be measured if two nuclei are close to each other, a statement which is sufficient for most applications. In the case of proton resonance, to which the descriptions given above refer, a maximum distance of 0.5 nm between two nuclei has proven to be a generally accepted limit if a NOE in proteins is detected. This is also in good agreement with experimental data and simulation calculations.

The determination of the distance using the nuclear Overhauser effect is based on the dependence of the dipolar interaction between two nuclear spins on their separation. However, the dipolar interaction exists not only between two nuclear spins, but also between one nuclear and one electronic spin. The latter interaction likewise depends on the separation and can therefore in principle be used for measuring the distance. Because of the much larger magnetogyric ratio of the electron, it is much stronger than that between two nuclear spins and should therefore be much more easily detectable. Hence, before the arrival of 2D NMR it was hoped that this dependence on the distance could be used for a quantitative determination of the distances in proteins.

However, proteins as such are diamagnetic, that is to say, they do not possess an unpaired electron spin. Consequently, in order to exploit paramagnetic effects, paramagnetic centers in the protein have to be created. The simplest method for doing this is the interaction with other atoms or molecules that possess unpaired electrons and are therefore paramagnetic. Particularly suitable for such interactions are organic free radicals and paramagnetic metal ions. The most favorable case in principle is if a paramagnetic substance is already bound to the protein under physiological conditions as cofactor or substrate, because these substances do not induce structural artifacts. If this ideal possibility does not exist, diamagnetic ligands can be replaced by paramagnetic ligands that are as similar as possible in their chemical and chemical-physical properties. Less attractive because more difficult to interpret are unspecific interactions of paramagnetic substances with the protein or the modification of the protein by covalently bound organic free radicals. If paramagnetic probes are unspecifically bound to proteins, there is the problem that the precise spatial arrangement of the paramagnetic group(s) with respect to the protein is not known. When using the relatively large spin label molecules, on the other hand, one can never be sure that they do not induce unexpected structural changes of the protein that prevent the determination of the native structures, the real aim of the experiment.

Paramagnetic substances have two qualitatively different influences on neighboring spins. In addition to the chemical shift, they can cause an additional paramagnetic shift of the resonance frequency of the observed nuclei and they can shorten the relaxation times.

The paramagnetic shift decreases with the third power of the distance r_{AS}, which means it is proportional to r_{AS}^{-3}, if it is only due to the dipolar coupling between the nuclear spin A observed and the unpaired electronic spin S. Since the paramagnetic relaxation rate is proportional to the square of the interaction (see Sect. 1.3), it decreases to first approximation with the sixth power of the distance r_{AS} and is thus proportional to r_{AS}^{-6} as we have already seen in the case of the NOE.

Since the electronic spins relax in general much faster than the nuclear spins, the correlation times relevant for the nuclear relaxation rates are frequently determined by the electronic relaxation time. In this range of correlation times the dipolar relaxation rate becomes smaller, the shorter the correlation time. If the electronic spin relaxation time T_{1e} is very short, the line broadening by transverse relaxation becomes small and the paramagnetic shift is the predominant effect (shift reagent). Inversely, in the case of T_{1e} values longer than about 0.1 ns, the line broadening is usually so large that the paramagnetic shifts become undetectable (Table 3.4).

While the nuclear spins are strictly localized, electronic spins are in general not. In organic free radicals, for instance, the electronic spin density is often distributed over large parts of the covalent or conjugated molecular frame, and even with non-covalently bound metal ions a non-vanishing spin density is frequently observed at the nearest ligands. Since the electronic spin density at the site of the nucleus often does not vanish, the so-called Fermi interaction may occur in addition to the dipolar interaction; hence, the paramagnetic shift δ_{hf} due to this hyperfine interaction can be divided into two parts, the Fermi contact part δ_{Fc} and the purely dipolar part as discussed in Sect. 1.2.4. The dipolar interaction, the effect of which is classically described by the local magnetic fields of the interacting magnetic moments, leads to a change of the resonance frequencies that in

Table 3.4. Relaxation times and line broadening of important paramagnetic substances[a]

Substance	T_{1e}/ns	Line broadening/Hz
Cu^{2+}	1–3	3000–9000
Organic free radicals	>10	>20000
VO^{2+}	1–10	3000–20000
Mn^{2+}	1–10	40000–200000
Fe^{3+} ($S = \frac{5}{2}$)	0.01–0,1	400–5000
Gd^{3+}	1–10	60000–400000
Ni^{2+}	0.001–0,1	25–1000
Fe^{2+} ($S = 0$)	0.001	70
Co^{2+} ($S = \frac{3}{2}$)	0.0001–0,01	50–200
Fe^{3+} ($S = \frac{1}{2}$)	0.0001–0,01	10–40
Dy^{3+}	0.001	100
Ho^{3+}	0.001	100
Yb^{3+}	0.001	30

[a] Values at B_0 = 2.35 T, r_{AS} = 0.5 nm (according to Bertini et al., 1987)

Table 3.5. Pseudo-contact shift by lanthanides[a]

Lanthanide-ion	Theoretical pseudo-contact shift (ppm)
Ce^{3+}	6.3
Dy^{3+}	100
Er^{3+}	−33
Eu^{3+}	−4.0
Gd^{3+}	0
Ho^{3+}	39
La^{3+}	0
Lu^{3+}	0
Nd^{3+}	4.2
Pm^{3+}	−2.0
Pr^{3+}	11.0
Sm^{3+}	0.7
Tb^{3+}	8.6
Tm^{3+}	−53
Yb^{3+}	−22

[a] The pseudo-contact shifts are referred to the effect of Dy^{3+} (according to Bleaney et al., 1972)

the simplest case depends on the angle between the line connecting the two nuclei and the external magnetic field. However, in solution this effect is averaged out by rotational diffusion unless the magnetic susceptibility of the paramagnetic center is anisotropic, that is, it possesses a dependence on the orientation. The anisotropy is an additional requirement for a substance to be used as a shift reagent. A sufficient anisotropy is usually not found with organic free radicals, but occurs with rare earth and several transition metal ions (Tables 3.4 and 3.5).

Their influence on the chemical shift is usually termed the pseudo contact shift δ_{pc}. The chemical shift δ observed is then the combination of the diamagnetic part δ_{dm} and the hyperfine part δ_{hf}

$$\delta = \delta_{dm} + \delta_{hf} = \delta_{dm} + \delta_{Fc} + \delta_{pc} \quad . \tag{3.13}$$

Information concerning the distances is not easily derived from the Fermi contact shift, but it provides a considerable enlargement of the range over which the resonance frequencies are distributed. The resulting increase of the resolution often permits an identification of individual resonance lines in very large proteins. If a Fermi contact shift is observed, this provides simultaneously the additional information that a common frame of bonds exists between the paramagnetic center and the observed nucleus.

If one assumes for simplicity that the spin density is centered at one metal ion, the pseudo contact shift in ppm is given by

$$\delta_{pc} = \frac{F(\theta, \varphi)}{r_{AS}^3} \quad . \tag{3.14}$$

Here r_{AS}, θ and φ determine the position of the spin A observed with respect to a molecular coordinate system centered in the paramagnetic center S (Fig. 3.17). Hence, the pseudo contact shift decreases with the third power of the distance r_{AS}^{-3} between the paramagnetic center S and the nuclear spin A, but in addition it depends on the angles θ and φ. $F(\theta, \varphi)$ is given by

$$\begin{aligned} F(\theta, \varphi) = -\frac{10^6}{N_A} \Big\{ &\frac{1}{3} \Big[\chi_z - \frac{1}{2}(\chi_x + \chi_y) \Big] (3\cos^2\theta - 1) \\ &+ \frac{1}{2}(\chi_x - \chi_y)\sin^2\theta \cos^2 2\varphi \Big\} \quad . \end{aligned} \tag{3.15}$$

The size of the chemical shift depends on the size of the principal values χ_x, χ_y and χ_z of the molar magnetic susceptibility which is characteristic for each para-

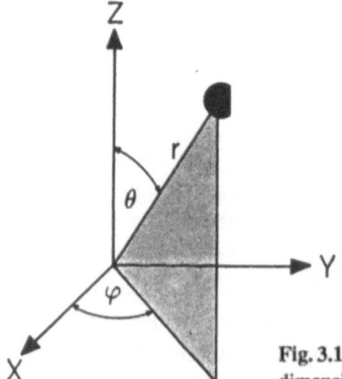

Fig. 3.17. Polar coordinates, description of the position in the three-dimensional space by the coordinates r, θ, φ

magnetic center in a given complex (N_A = Avogadro's number). In the frequently occurring case that the paramagnetic center possesses simple axial symmetry, χ_x equals χ_y and the dependence on the angle φ vanishes. Although appropriate theoretical expressions exist for the various paramagnetic centers that connect χ with observable quantities, the main problem is the precise determination of the components of χ in each individual case. However, if the positions of several resonance lines with respect to the paramagnetic center are well known, they can be used for calibration and then for structure determination.

An important prosthetic group that is built into many proteins and shows numerous paramagnetic interactions is porphyrin and its derivatives. Porphyrin forms chelate complexes with metal ions, like iron, magnesium, zinc, nickel, cobalt, copper and rare earths. An important complex is the iron complex, the heme, that plays an important role not only in hemoglobin for binding oxygen, but also in a number of other reactions, such as the electron transport of the cytochromes, H_2O_2 conversion in catalysis or the oxidation of fatty acids by peroxidases. In these complexes the iron ion can either be diamagnetic or paramagnetic depending on the state of oxidation or the type of ligands. Hence, one often distinguishes natural diamagnetic and paramagnetic complexes of the same molecule, the NMR properties of which are suitable for comparison. Furthermore, metal ions other than iron can also be introduced into the ring system which facilitates considerably the interpretation of the NMR results, for instance, by separation of contact and pseudo contact shifts. Besides the paramagnetism centered at the metal ion in the heme, additional porphyrin radicals exist in which the unpaired spin is distributed over the whole porphyrin frame. The resulting chemical shifts in such systems can be considerable. Instead of chemical shifts between 0 ppm and 10 ppm normally obtained with the ^{1}H-NMR method, in this case frequently shifts within the range of about ±100 ppm can be observed. Figure 3.18 shows such a ^{1}H-NMR spectrum of the horse-radish peroxidase that contains a heme as prosthetic group.

Metals of the group of lanthanides are particularly suitable as substitutes for the physiologically occurring divalent ions Mg^{2+} and Ca^{2+} at their sites in enzymes. With the exception of the diamagnetic ions La^{3+} and Lu^{3+} and of Gd^{3+}, which leads to a line broadening, the other twelve lanthanides are typical shift reagents. The metal ions often exchange sufficiently fast with the surrounding solvent to satisfy the condition of fast exchange. The observed chemical shift of the proton resonance is then averaged again over the protein population with and without metal ions as we have already seen in Sect. 2.2.2. Parvalbumin is a representative of a class of calcium binding proteins with a high affinity for calcium, and their sites for binding calcium possess a very similar structure.

Figure 3.19 shows the effect that occurs if the diamagnetic calcium ions are replaced with paramagnetic Yb^{3+} ions. The hyperfine shifts observed are not as large as in the example shown above, but still rather marked. An X-ray structure of the carp parvalbumin exists, which can be used for a first analysis of the position of the coordinate system centered at the metal ion with respect to the

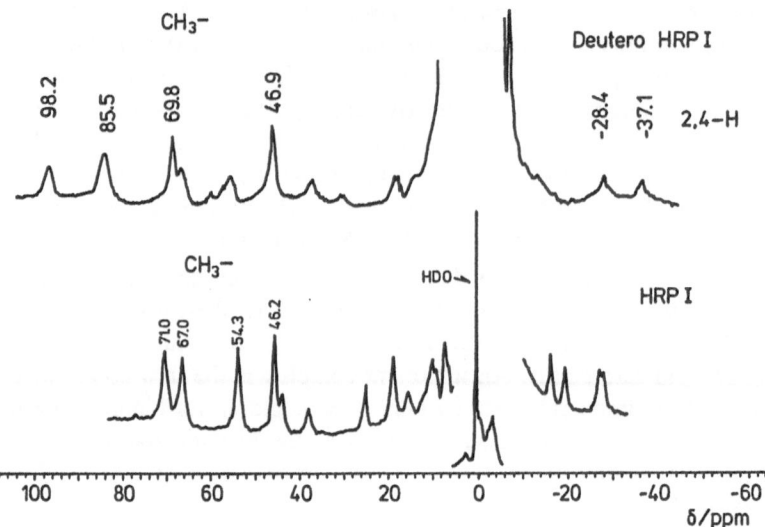

Fig. 3.18. 300 MHz ^{1}H-NMR spectrum of the horse-radish peroxidase that contains a heme as prosthetic group. The naturally occurring heme was replaced by a heme derivative in the upper spectrum (Morishima et al., 1986, with permission)

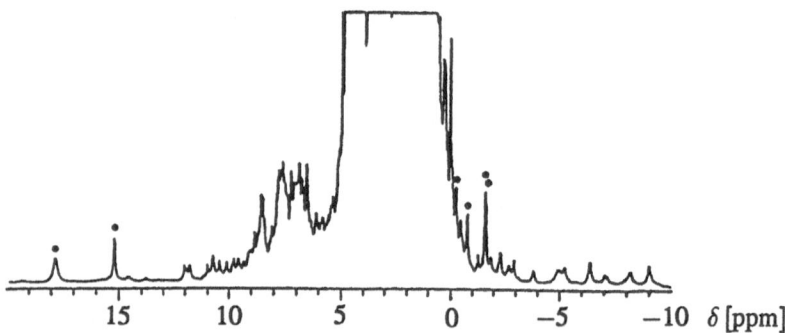

Fig. 3.19. Lanthanide-induced shift in a calcium binding protein. 270 MHz ^{1}H-NMR spectrum of parvalbumin that has bound Yb^{3+} ions instead of the naturally occurring Ca^{2+} ions. The central part of the spectrum containing the majority of all protein resonances and the water resonance is represented incompletely. (●) labels the methyl resonances of the protein that are shifted strongly by the paramagnetic ion (Lee and Sykes, 1983, with permission)

protein, and can serve for an estimation of the size of the components of χ in (3.15). Using these parameters one can calculate the surface shown in Fig. 3.20 that would produce a pseudo contact shift of 27 ppm for a proton situated on this surface. The shape of this surface is given again by equations (3.14) and (3.15).

Analogous to the shift effect, the effect of paramagnetic substances on the relaxation time consists of two parts, the Fermi contact part and the dipolar part. Since it is here a matter of influence on relaxation times, the interaction enters quadratically as always and the dipolar part is proportional to r_{AS}^{-6}. If one wants to deduce distances from the change of the relaxation times, it is in general assumed

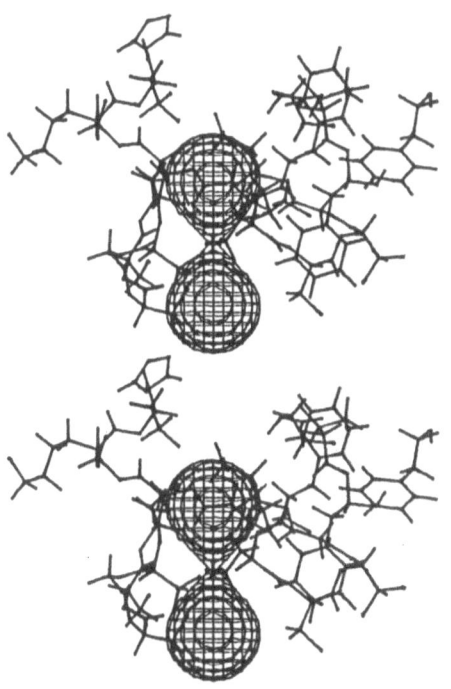

Fig. 3.20. The position of the Yb^{3+} ions in parvalbumin. The plane is shown where nuclear spins near the Ytterbium ion would experience a paramagnetic shift of 27 ppm (Lee and Sykes, 1983, with permission). The metal ion is located in the center of symmetry of this surface structure. Representation in stereo view

that the Fermi contact part is negligible, a condition which is usually rather well satisfied if no covalent bonds exist between the paramagnetic center and the nucleus observed, and if the ligand is not situated in the immediate neighborhood of the paramagnetic center. The longitudinal and transverse relaxation rates in the paramagnetic complexes $1/T_{1M}$ and $1/T_{2M}$ can then again be divided into a paramagnetic part $1/T_{1,2P}$, which is caused by the unpaired electrons of the paramagnetic center, and a diamagnetic part $1/T_{1,2D}$, which would be present if other factors remain unchanged but there where no unpaired electrons. The latter condition can be simulated approximately if the paramagnetic ligand is replaced with a diamagnetic one that is otherwise as similar as possible. In the case of paramagnetic substances with comparatively long relaxation times like Mn^{2+} and Gd^{3+} (Table 3.4) the effect on the transverse relaxation time is so strong that the linewidths of the resonance lines of nuclear spins that are fairly close to the paramagnetic center (< 1.5 nm) are so drastically broadened that they become invisible in the spectrum.

For determining smaller distances the averaging by fast exchange can be useful, as is frequently found with non-covalently bound ligands. The relaxation rates $1/T_{1,2obs}$ are then equal to the weighted average of the relaxation rates observed in the free state $1/T_{1,2f}$ and in the bound state $1/T_{1,2M}$:

$$\frac{1}{T_{1,2obs}} = \frac{p_f}{T_{1,2f}} + \frac{p_b}{T_{1,2M}} \quad . \tag{3.16}$$

The averaging is then performed over the occupation probability in the free state p_f and in the bound state p_b ($p_f + p_b = 1$). These simple relations are valid if the average lifetime in the paramagnetic complex τ_M is markedly shorter than the relaxation times $T_{1,2M}$ and the inverse of the difference of the chemical shifts in the free and in the bound states $1/|\Delta\delta|$. By suitably choosing the experimental conditions the relaxation times must be set in such a manner that they result in a well-resolved observable linewidth.

In addition to paramagnetism, another effect exists, the diamagnetic anisotropy, which depends on the distance and can therefore in principle be used for structure determination. An effect that is at least qualitatively understandable is the ring current shift, which can be observed in aromatic ring systems. Depending on the position of the nuclear spins observed with respect to the ring system, it can lead to an increase or a decrease in the resonance frequency. A simple plausible explanation is that a current is induced in the π-orbitals by the external magnetic field analogous to a conducting loop; its magnetic field weakens the external magnetic field inside the conducting loop (Pauling, 1936). Although this simple idea turned out to be sufficient for a qualitative description, for a long time the question of whether or not a ring current exists remained a controversial issue.

Four important models exist for describing the diamagnetic anisotropy of aromatic ring systems: the classical dipolar equation that approximates the induced magnetic field by a magnetic dipole in the center of the ring system (Pople, 1956), the double dipole equation in which the magnetic field is approximated by two dipoles above and below the plane of the ring system (Abraham et al., 1977), the semi-classical Johnson-Bovey equation (Johnson and Bovey, 1958) where the magnetic field of two current loops above and below the plane of the ring system is calculated, and the quantum mechanical Haigh-Mallion equation (Haigh and Mallion, 1972). Studies of proteins whose X-ray structures are known show that for nuclei which do not belong to the aromatic spin system all models can, after an appropriate calibration, reproduce the ring current shift to a good approximation. This is particularly true for simple ring systems, as in tyrosine, phenylalanine and histidine. For more complex ring systems as in tryptophan and the heme the Johnson-Bovey equation seems to give the best results. The simplest ring current equation is the dipolar equation in which the ring current shift δ_R (in ppm) depends on a constant C for the ring system observed, and also on the distance r from the ring center and on the angle θ (r and θ are as defined in Fig. 3.17 if the z-axis is perpendicular to the plane of the ring):

$$\delta_R = -C\frac{3\cos^2\theta - 1}{r^3} \quad .$$

(3.17)

The constant C is given by:

$$C = \frac{i\chi_L \times 10^6}{3N_A}$$

(3.18)

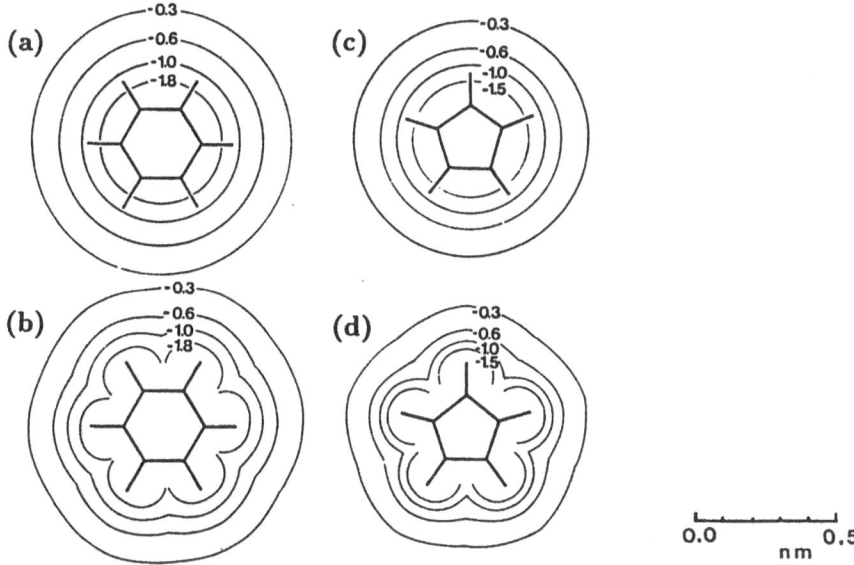

Fig. 3.21a–d. The diamagnetic anisotropy in aromatic ring systems. The change of chemical shift δ_R (in ppm) expected in the neighborhood of the aromatic rings in the ring plane. (a) and (c) were calculated with the Johnson-Bovey equation, (b) and (d) with the Haigh-Mallion equation (Perkins and Dwek, 1980, with permission). With our definition, δ_R would have the opposite sign

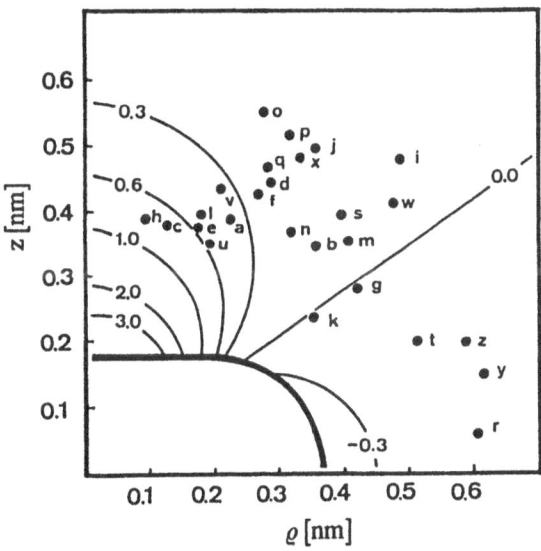

Fig. 3.22. Ring current shifts in the plane perpendicular to the ring system through the ring center. The shift δ_R (in ppm) was calculated with the Johnson-Bovey equation. The solid line marks the size of the aromatic ring, the data points shown were obtained from methyl groups in lysozyme (Perkins and Dwek, 1980, with permission). (Sign of δ_R see Fig. 3.21)

with a ring current intensity factor i ($i = 1$ for phenylalanine), Avogadro's number N_A and the molar susceptibility $\chi_L = 49.5 \cdot 10^{-12} \mathrm{m}^3 \mathrm{mol}^{-1}$.

Figure 3.21 shows the ring current shifts expected for hexagonal and pentagonal rings when the Johnson-Bovey and the Haigh-Mallion equations are applied. For larger distances r the two equations give a rather similar result. Figure 3.22 shows the ring current shift in a plane through the ring center perpendicular to the plane of the ring. A difficulty when exploiting the ring current shift for structure calculations is the existence of other mechanisms leading to conformation-dependent changes of the chemical shift. Since these influences are difficult to separate, the value of δ_R can only be predicted with relatively low precision.

3.1.4 Assignment of Resonance Lines

One of the most important preconditions for a reliable structure determination is in most cases the assignment of the NMR resonance lines to particular atoms in the macromolecules. Although in principle the assignment of a few resonance lines suffices for obtaining some information on the structure, the accuracy of the structural information increases with the number of lines assigned because this increases the number of possible consistency tests as well. The reliability of the structural information increases considerably with the accuracy of the assignments because it is frequently very sensitive with respect to erroneous assignments. When assigning the resonance lines of ^{1}H-NMR spectra of proteins, one usually proceeds in two steps: one tries first to determine which resonance lines belong to one amino acid residue neglecting the position of this amino acid in the protein sequence. When doing this, in many cases information on the type of amino acid is obtained as well. The next step is the fitting into the protein sequence. The sequence is in general known from independent methods, either directly from amino acid analysis or indirectly from sequencing of the corresponding DNA. Hence, the assignment to a specific location in the sequence depends on the accuracy of the primary structure, although smaller mistakes in the primary structure can occasionally be detected by inconsistencies in the NMR data. This assignment is based on relations between the distances that are typical for amino acids following each other in the sequence. Simultaneously these relations between the distances are also characteristic for the positions in certain secondary structure elements, so that with the sequential assignment information on the secondary structure may also be obtained.

As stated above, the assignment of the resonance lines of certain types of amino acids would be simple if the resonance frequencies were a simple function of the type of the amino acid and of the position of the nucleus in the amino acid residue. However, owing to the conformation-dependent changes of the chemical shift the presence of a certain amino acid cannot be deduced directly from the ^{1}H-NMR resonance frequency, but only with a certain probability (Fig. 3.23).

In addition to two-dimensional NMR there exist two generally applicable biochemical methods, the selective deuteration of certain amino acids and the

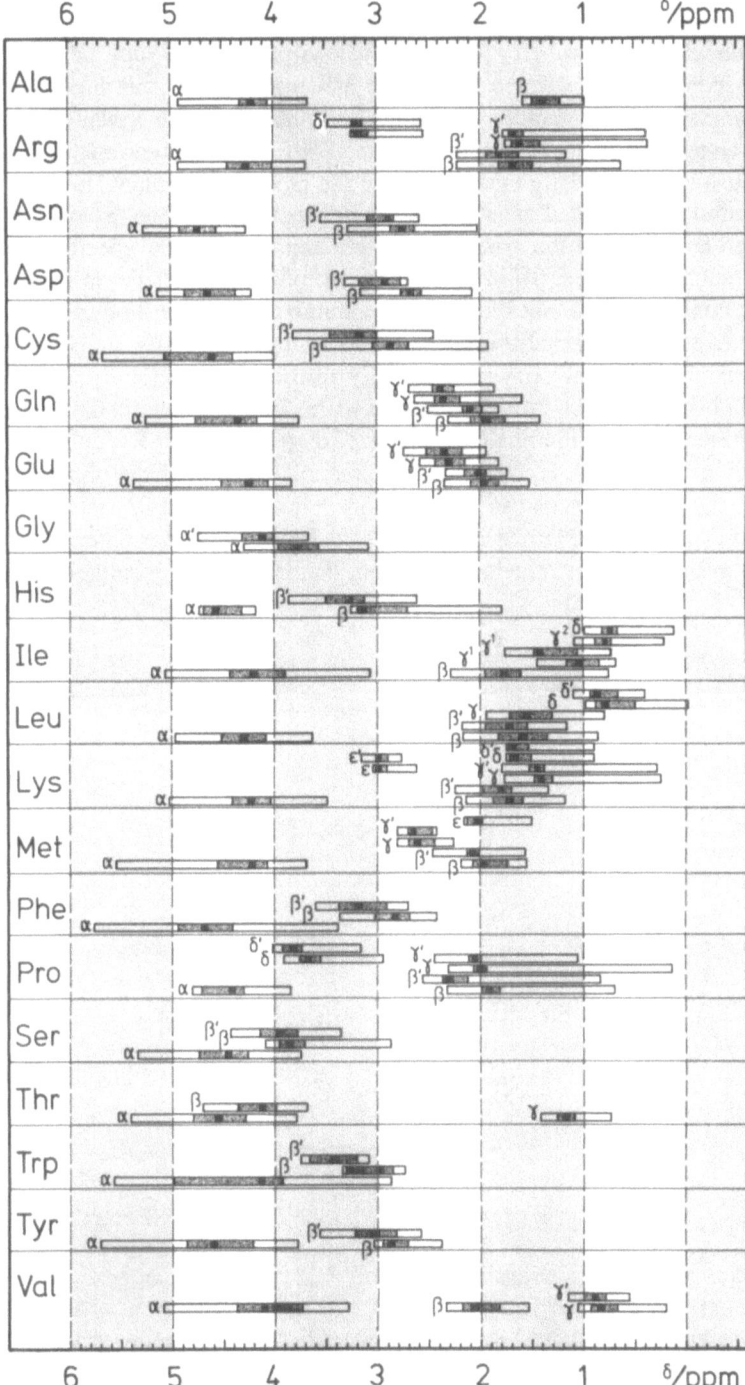

Fig. 3.23. Distribution of chemical shifts δ (referred to the internal standard DSS) in ^{1}H-NMR spectra of proteins. The medians, the quartiles, the 5th and 95th percentiles of the distribution of chemical shifts are given (Gross and Kalbitzer, 1988, with permission)

comparison with almost homologous proteins where only a few amino acids are exchanged in the sequence. Although both methods were already successfully applied before the breakthrough of two-dimensional NMR, it is only now that they have become really effective due to the modern developments of molecular biology. Selective deuteration is usually performed in such a way that a certain deuterated amino acid is supplied to the cells in the culture medium in the hope that it will be directly incorporated in the protein concerned. However, since the amino acids offered are not directly incorporated into the protein, but are converted and used for the synthesis of other amino acids, the selectivity of the incorporation is reduced. To increase the selectivity it is possible to choose auxotrophic mutants that cannot synthesize the amino acids in question by themselves and so use the externally offered amino acid primarily for the peptide synthesis. Initially this method was essentially restricted to bacterial enzymes because only bacteria can be grown on a large scale. To extend these methods to proteins of higher species, in principle cultures of corresponding cells could have been used, a method which is, however, not practical because of the high costs involved. Today these proteins can also be cloned and expressed in bacteria, so this problem is now solved. Furthermore, one can choose conditions under which the bacteria overexpress the protein in question, that is, produce essentially only

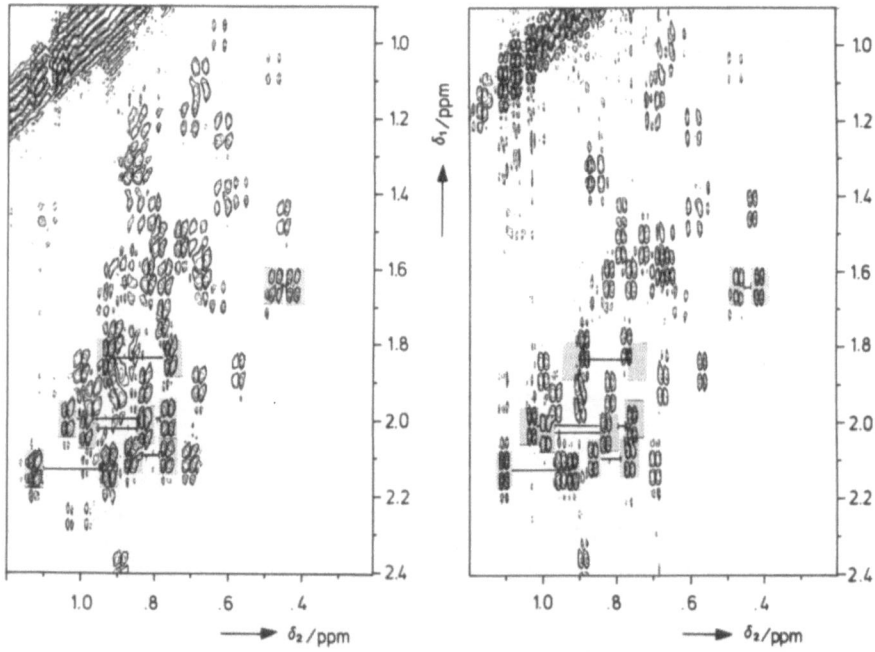

Fig. 3.24. Double quantum filtered COSY spectra (500 MHz) of HPr protein from *S. aureus* (*left*) and *S. carnosus* (*right*). The Hβ-Hγ3 cross peaks of the 6 and 5 valine residues are marked. One pair of cross peaks at $\delta_1 = 1.83$ ppm is missing in the spectrum of HPr protein from *S. carnosus*

this protein. This, of course, renders the use of the rather expensive deuterated amino acids much more effective.

Interpreting the spectra obtained is simple. If, for instance, deuterated leucine was incorporated, its deuterons no longer give ^{1}H-NMR signals. Hence, all vanishing resonance lines compared with the normal spectrum are due to leucine. Comparison of proteins with very similar sequence renders similar information. The vanishing resonance lines belong to the amino acids exchanged. In this case one can again use either natural mutants or new mutants synthesized on purpose. Figure 3.24 shows an example for the comparison of natural homologues, the HPr proteins of *Staphylococcus aureus* and *Staphylococcus carnosus*. Both differ only in two positions in the sequence: HPr of *Staphylococcus aureus* contains in position 7 a valine and in position 71 a serine, while *Staphylococcus carnosus* contains at these positions a threonine and an alanine. In the resolution of the two-dimensional spectroscopy it is easy to see that the corresponding cross peaks in the spectrum of HPr of *Staphylococcus carnosus* have vanished. If the signals of only one single amino acid have disappeared by one of the two methods, an additional assignment to the positions in the sequence is obtained. With a well-defined mutagenesis this condition can always be satisfied, in contrast to deuteration when all amino acids of one type vanish unselectively.

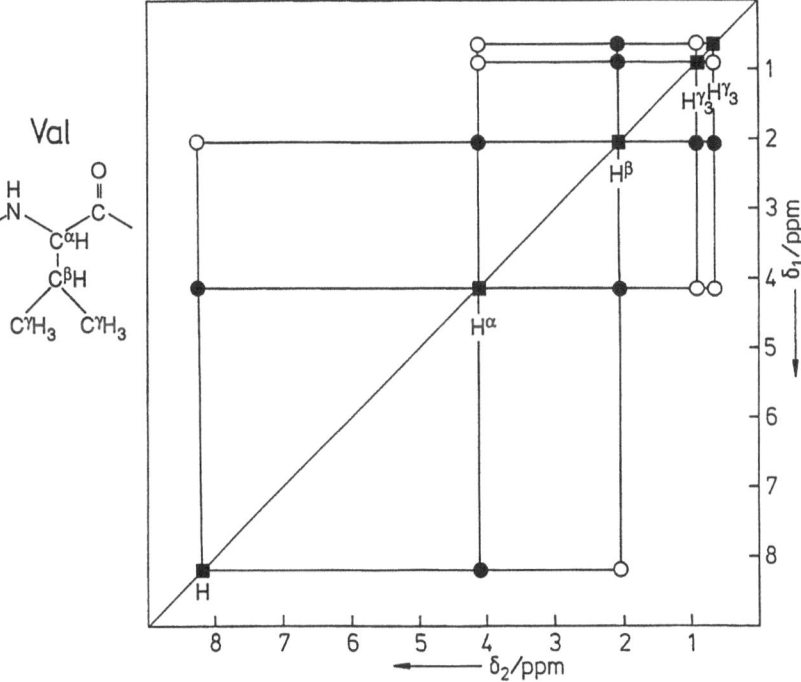

Fig. 3.25. Schematic view of a coupling pattern in the COSY and RCT spectra of a valine residue. (■) diagonal peaks, (•) COSY cross peaks, (o) RCT cross peaks. Usually, COSY cross peaks are also observable in an RCT spectrum. The multiplet structure of cross peaks is not shown

In principle the more elegant method is the exclusive application of 2D-NMR methods for a complete assignment of all resonance lines; this fails, however, for excessively large molecular weight.

The assignment of the resonance lines of certain types of amino acids is based on the fact that in the amino acid residues most protons are J-coupled to each other, while between protons of two neighboring amino acids no measurable J-coupling exists because between the closest pair of protons, the $H\alpha$ and the amide proton, there are four chemical bonds (Fig. 3.3). Hence, in this sense each amino acid residue of the protein can be considered as a separate spin system. Thus one obtains a typical coupling pattern for each amino acid that can then be determined with 2D-spectroscopy. Figure 3.25 shows schematically the coupling pattern of a valine residue as it is expected in a COSY- and a RCT-spectrum.

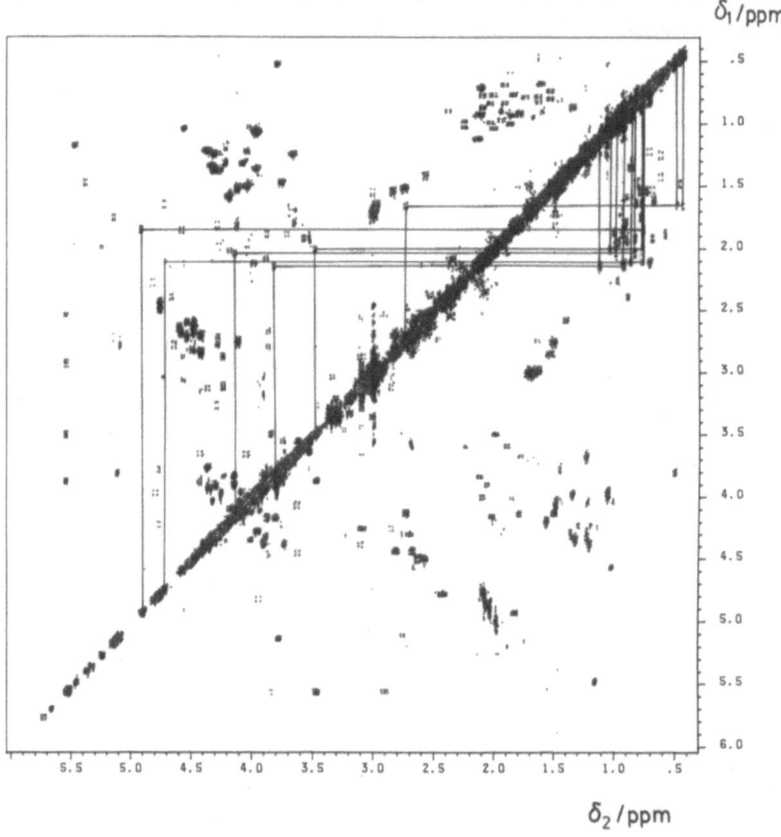

Fig. 3.26. Valine patterns in the double quantum filtered COSY spectrum (500 MHz) of the HPr protein from *S. aureus* in D_2O. The valine patterns are depicted. The spectrum was recorded in approximately two days, protein concentration 4 mM

This pattern must then reappear in the real spectrum (Fig. 3.26). The difficulty lies not only in the fact that the precise chemical shift of the resonances looked for is unknown, but also in that the individual cross peaks cannot be found with certainty in the spectrum. This can be due to the large linewidth of the cross peak; since the amplitude decreases accordingly, it can no longer be distinguished from the background. Since the linewidth, which is essentially determined by the transverse relaxation and the chemical exchange, is not known *a priori*, there is no guarantee that an expected cross peak can really be detected. The linewidth has a particularly strong influence on the 2D-spectra of the COSY type; the splitting of cross peaks into subpeaks with positive and negative signs is such that the total integral over the subpeaks of a cross peak multiplet is zero. The more the width of the individual lines increases, the more the lines cancel each other. The same happens when lines with different signs come closer to each other, which is actually the case when the coupling constant between the spins to which the cross peak belongs becomes smaller and smaller until they vanish at $J = 0$. Since the coupling constants depend on the dihedral angle they are conformation-dependent, and it may happen that individual cross peaks vanish because of this effect.

A second problem is the superposition of the cross peak looked for with other signals, which can prevent their unambiguous identification in the spectrum. In this case the transition to another 2D-method may be helpful. For example, if the identification of the cross peaks belonging to the spin system is difficult in a COSY spectrum because of superposition effects, it often might be possible in an RCT or TOCSY spectrum because the corresponding patterns of the cross peaks differ in both experiments and hence often appear at other, superposition-free positions in the spectrum.

Beside the biochemical methods mentioned above, there still exist a number of other methods for a specific modification of the side chains in the intact molecule that allow the correlation of the spectral changes of amino acids in the sequence. Very often these methods take advantage of the fact that in the native state of the protein the reactivity of the same amino acid can vary greatly with the position in the molecule. For example, an amino acid in the inner part of the protein sometimes cannot be reached by a chemical reagent. The position of the modification in the sequence has then to be determined with chemical methods. Usual methods are the H-D exchange of the histidine $\varepsilon 1$-protons or the nitration of tyrosines. However, with all these methods there is the risk that the spectral changes are incorrectly interpreted if they do not result from the specific labelling assumed, but from unspecific denaturation processes.

The most elegant method is again 2D-NMR spectroscopy. It is based on the fact that spin systems can be fitted into the sequence if the type of the amino acids is known and if it is also known which of these spin systems belong to the neighboring amino acids in the sequence. If the neighbors of all amino acids are known, the sequence of the protein is determined with NMR. For the assignment it usually suffices to know only a few neighbors because with this information parts of the sequence can be unambiguously incorporated into the known sequence. In the simplest case, if an amino acid occurs only once in the protein, no neighboring systems have to be known for the incorporation into the sequence.

How do we obtain this information on the neighbors? Since the J-coupling does not extend beyond the peptide bond, it cannot serve as a source of information. There remains the nuclear Overhauser effect that can be observed between protons of neighboring amino acids if they come sufficiently close to each other. Figures 3.3 and 3.4 show that this is a reasonable assumption. However, the distances depend on the dihedral angles. The most important distances are plotted in Fig. 3.27 using the most common nomenclature. Apart from the proline residue that does not contain an amide proton, computer simulations show that for any allowed values of the φ- and ψ-angles always at least one of the distances d_{NN}, $d_{\alpha N}$, and $d_{\beta N}$ is smaller than or equal to 0.3 nm, that is, small enough to secure a measurable nuclear Overhauser effect in the normal case. In principle the protons may, of course, come close to each other in the 3D-structure even if they do not belong to neighboring amino acids. However, this is not very likely as was revealed by a statistical analysis of protein crystal structures: 88% of all distances $d_{\alpha N}$ and d_{NN} as well as 76% of all distances $d_{\beta N}$ that are 0.3 nm or less in these protein structures belong to neighboring amino acids and would therefore provide correct sequential NOEs (Billeter et al., 1982). The ambiguity that remains can again be eliminated by comparison with the existing amino sequence.

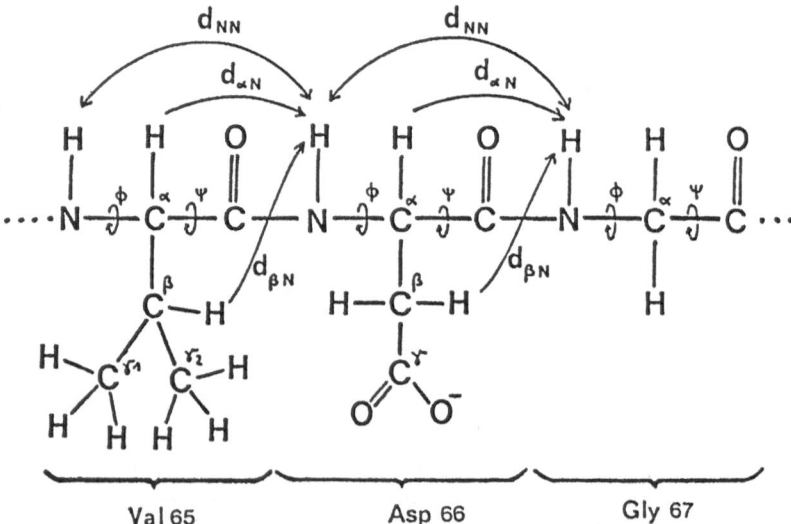

Fig. 3.27. Sequential NOEs. The most important distances between amino acids that lead to easily observable NOEs in amino acids which are neighbors in the sequence. The amino acids shown are part of the primary structure of HPr from *S. faecalis*

The sequential distances d_{NN} and $d_{\alpha N}$ depend on the dihedral angles φ and ψ of the main chain, the size of which determines, by definition, the secondary structure. Hence, the sizes of the corresponding NOEs are also characteristic of the secondary structures present.

3.1.5 Pattern Recognition in Two-Dimensional NMR Spectra

The non-computer-aided interpretation of two-dimensional ^{1}H-NMR spectra of biological macromolecules is time consuming and rather dull. For a small protein, such as the HPr protein of *Staphylococcus aureus* with 87 amino acids, one expects about 600 cross peak multiplets in the COSY spectrum and about 1000 in the RCT spectrum. In these spectra, first the coordinates of the cross peaks must be measured and then the patterns of the amino acids identified from among the multitude of all cross peaks. After this is done, the NOESY spectra have to be evaluated to obtain the sequential assignment. Usually not only one set of spectra measured in H_2O is evaluated, but the same patterns are looked for in a set of spectra recorded in D_2O. The latter spectra are of better quality because the water signal is weaker. However, they do not contain information on the protons which exchange with the solvent. It is obvious that this data evaluation is the main bottleneck in a routine application of 2D NMR for determining the structure of proteins. Hence, it is tempting to try to automatize these processes.

There are essentially two basically different ways to achieve a pattern recognition. In principle, the best procedure is to put all available information into the evaluation of the experimental spectra. The basic physical interactions are known that make it possible to simulate any 2D experiment if the spin system is well enough characterized; however, the relaxation phenomena are mostly described only in a simple approximation. The spin system itself is known from the chemical structure of the amino acids; the corresponding coupling constants and chemical shifts are the free parameters that have to be adapted to the experimental results. Their values describe the completely recognized patterns. In practice this procedure requires much computing. It can be somewhat improved if a set of special complementary 2D experiments is performed. Fig. 3.28 shows that by using this method a complete description can be obtained for small molecules even from very noisy spectra.

Alternatively, one can proceed stepwise as in the non-computer-aided evaluation and extract all important features from the 2D-spectra at a very early stage in order to be able to continue processing on a higher level of abstraction (Fig. 3.29). This method does not make use of the total information; on the other hand, it requires less computing and is also applicable to spectra of large molecules.

A problem for the automatic evaluation procedure involves artifacts in the spectrum. They are often easy to recognize with the eye. Their various features with all their possibilities of variation are difficult to incorporate into a computer program. To deal with all these difficulties, all possible experimental measures should be taken to keep them as low as possible.

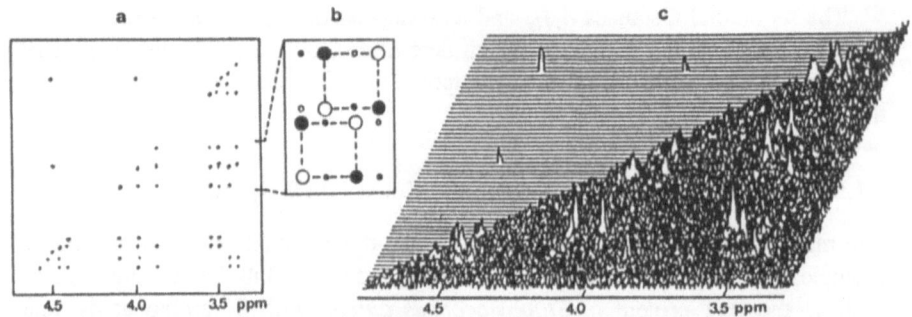

Fig. 3.28a–c. Pattern recognition in a simple 2D spectrum. Double quantum filtered COSY-45 spectrum of 2,3-dibrompropionic acid. (a) Patterns found, (b) partial patterns, (c) experimental spectrum (Meier et al., 1984, with permission)

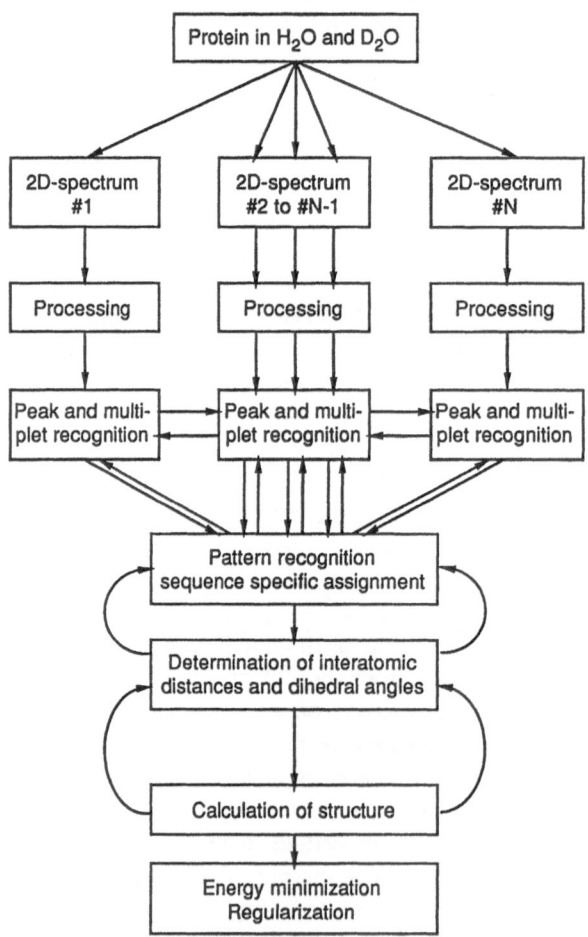

Fig. 3.29. Simplified scheme for pattern recognition in 2D NMR spectra

112

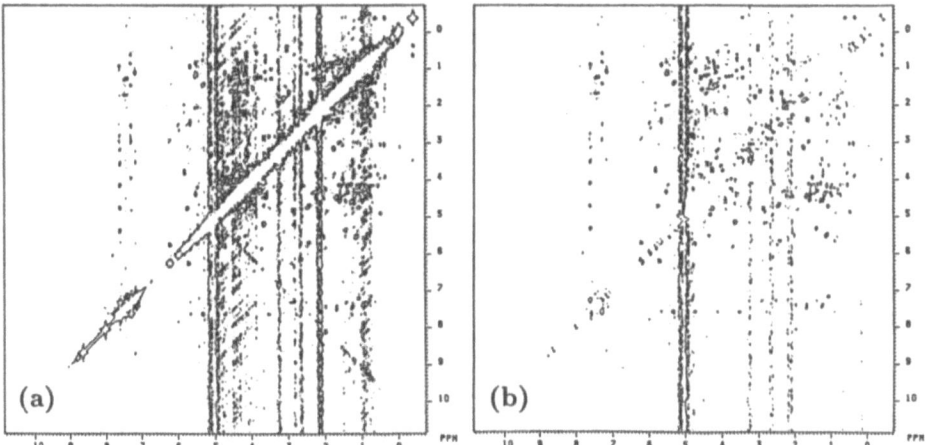

Fig. 3.30a,b. Reduction of artifacts in 2D spectra. 360 MHz NOESY spectra of HPr protein from S. faecalis (mixing time τ_m = 100 ms). (a) Experimental spectrum showing typical artifacts, t_1 noise (perpendicular stripes in the spectrum) and an antidiagonal (stripes from the left upper corner to the right bottom corner of the spectrum). (b) Processed spectrum where the artifacts were strongly suppressed and where the intensity of the diagonal was reduced (Glaser and Kalbitzer, 1986, with permission)

As a next step one has to try to further reduce these artifacts by suitably processing the 2D data matrix before proceeding with the pattern recognition (Fig. 3.30). In the NOESY spectrum shown there are two typical experimental artifacts, the antidiagonal (a mirror image of the normal spectrum) and the t_1 noise (the intense stripes that run vertically from top to bottom). Both artifacts can be reduced by suitable processing. Simultaneously the intensity of the diagonal peaks is reduced; these do not contain information used in the normal evaluation, but can superpose the cross peaks in their neighborhood.

On the next level, peak and pattern recognition, the main part of the remaining artifacts can be recognized and eliminated. Nevertheless, for practical applications it proved to be useful to check the results on different levels of the evaluation by an interactive control. Multi-dimensional NMR experiments promise to facilitate the computer-aided data evaluation because they are almost free of overlap and exhibit a high redundancy of information.

3.1.6 Structure Determination from NMR Data

After as complete as possible an assignment of the resonance lines, which as such is only of minor biological relevance, there follows the determination of the structure of the protein from the interatomic distances that can be derived from the NOEs as described above. Without major calculations, inferences concerning the secondary structure of the protein can be deduced from the existence of certain NOEs. A typical example are the sequential NOEs discussed earlier. If one assumes a simple r^{-6} dependence for each secondary structure element then the relative size of the NOEs to be expected can be calculated on the basis of the known standard coordinates. Typical combinations of the NOEs are then obtained for each type of secondary structure; these permit the formulation of qualitative hypotheses from the experimentally observed NOEs (Fig. 3.31). These structural hypotheses can in addition be supported by measuring the J-coupling constants between the amide and $H\alpha$ protons that again are determined by the dihedral angle φ. Quantitatively, the dependence of the coupling constant J on the dihedral angle $\theta(\theta = 60° - \varphi)$ can be well represented by the Karplus equation

$$K = A\cos^2\theta - B\cos\theta + C \qquad (3.19)$$

Here A, B and C are empirical constants, whose sizes depend on the molecule being investigated. This dependence is well satisfied for the dihedral angle φ in peptides by using the values 5.41, 1.27 and 2.17 for the constants A, B and C as shown by DeMarco et al. (1978) (Fig. 3.32). As can be easily seen from Fig. 3.32, the Karplus equation is quadratic and can have in general more than one solution. A further complication arises from the fast rotation around the single bond that produces an averaging of the coupling constants (fast exchange). However, for globular proteins the coupling constants are expected to reproduce the dihedral angle in the main chain and in the inner part of the protein comparatively well as here is no noticeable averaging over markedly different angles. In this way the α-helix and β-pleated sheets are comparatively easy to identify. Figure 3.33 shows such an example, a large antiparallel β-pleated sheet that was identified by NMR methods in the HPr protein of *E. coli* and *S. faecalis*. Although in this range only 9 out of 25 amino acids are identical, in both proteins the NOEs are characteristic

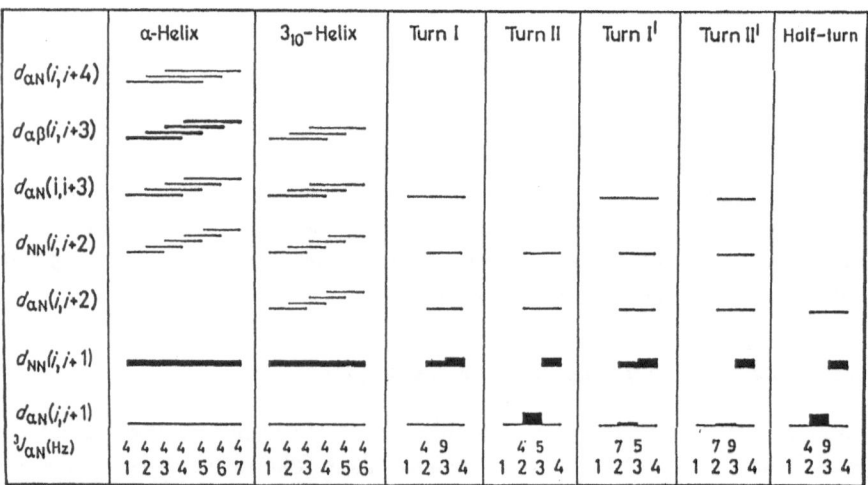

Fig. 3.31. NOEs and coupling constants $^3J_{\alpha N}$ that are expected for secondary structure elements in standard geometry. The lower row represents the relative position of the amino acid in the polypeptide chain, the indicated thickness of the lines is proportional to the strength of the NOEs expected (Wagner et al., 1986, with permission)

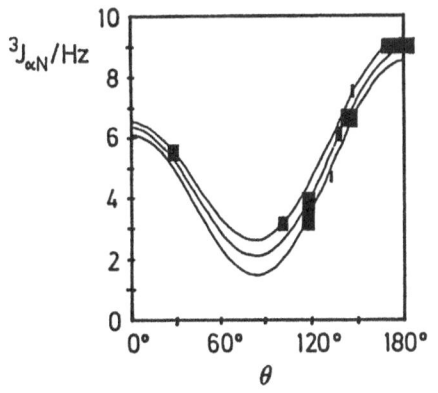

Fig. 3.32. Dependence of the vicinal proton-proton coupling constant $^3J_{\alpha N}$ on the dihedral angle φ in peptides. The curves shown represent the best fit of the Karplus equation (3.19) to the experimental data together with the error intervals (de Marco et al., 1978, with permission). The angle θ is defined by $\theta = 60° - \varphi$

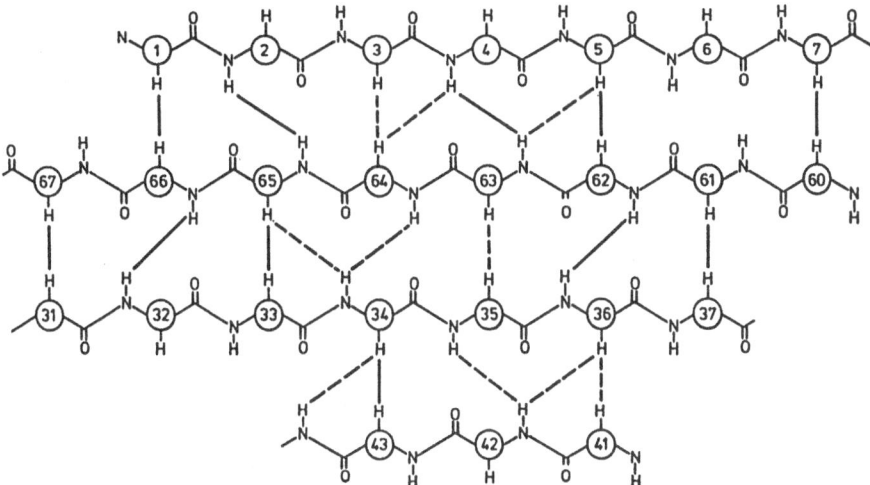

Fig. 3.33. The β-pleated sheet in HPr from *E. coli* (Klevit et al., 1986) and *S. faecalis* (Glaser, 1987) as resulting from NMR experiments. The solid lines represent the NOEs found in the two proteins, the broken lines are the NOEs found only in one of the proteins

of β-pleated sheets. This β-pleated sheet is also of interest for still another reason: it is one of the examples where the crystal structure differs fundamentally from the structure which could be determined with NMR in solution. In the crystal structure one finds two β-pleated sheets that are separated by a helix inbetween them (El-Kabbani et al., 1987).

Another often-cited example of these discrepancies was metallothionein-2, a metal binding protein. The binding pattern for the cadmium ions resulting from the X-ray structure analysis differed significantly from the one obtained by NMR investigations (Frey et al., 1985; Furey et al., 1986). However, a reinvestigation of the crystal structure revealed that NMR and X-ray structure are in fact almost identical (Stout, 1990). In principle, even large differences between NMR and X-ray structure do not mean that one of the two structures is incorrect, but rather that the crystal structure is not the predominating structure in solution. It can be predicted that in the future similar results will also be found with other proteins. This is particularly true for small proteins and polypeptides where the intermolecular interactions in the crystal are comparatively strong. It is therefore to be expected that NMR will become the method of choice for the structure determination of small proteins in solution.

To obtain a complete 3D structure of a protein from NMR data, one has to combine the experimentally measured interatomic distances with the possible conformations of the polypeptide chain in space and calculate structures that are in agreement with these restrictions. The methods which are mainly applied nowadays and provide satisfactory results are the distance algorithm (Braun et al., 1981; Havel et al., 1983) and restrained molecular dynamics (Kaptein et al., 1985).

The distance algorithm is an effective algorithm that is based in its original form on the description of the total structure in terms of interatomic distances. Possible degrees of freedom, such as the possibility of rotation around the dihedral angles, are taken into account by allowing a range between minimum and maximum distances. This method can easily be transferred to NMR data because the NOEs are particularly suitable for determining interatomic distances. The effectiveness of the distance algorithm has in the meantime been demonstrated in many cases. One possibility for testing this method is to construct artificial NOE data sets from X-ray structure data. For this purpose, first the distances have to be defined which correspond to expected observable NOEs. From the X-ray structure all proton pairs can be selected which come closer to each other than this maximum distance, that is, which in principle could provide an observable NOE. In order to obtain even more realistic conditions, the intensities to be expected can be divided into classes; one can, for instance, define strong, medium and weak NOEs corresponding to a distance of less than 0.25 nm, between 0.25 nm and 0.3 nm, and between 0.3 and 0.4 nm, respectively. Furthermore, the restricted quality of the experimental data may be simulated by not using all NOEs, but only a small selection for structure determination.

The crystal structure of BPTI is one of the best resolved of all structures and hence was chosen for such a test by Havel and Wüthrich (1985). Figure 3.34 shows the result. The thick line shows the position of the $C\alpha$ atoms in the crystal structure, the thin lines various calculated structures that agree with the given NOEs. For BPTI one obtains a total of 508 NOEs which are divided into various classes as defined above. Considering 365 NOEs for the calculation, one still obtains an excellent agreement between the original structure and the calculated structure (Fig. 3.34a). Even if the calculation is restricted to the 170 strongest NOEs, the global folding of the protein is still rather well defined (Fig. 3.34b). A further proof of the reliability of NMR was provided by two groups, an X-ray structure group (Pflugrath et al., 1986) and an NMR group (Kline et al., 1986) which independently investigated the structure of the α-amylase inhibitor tendamistat that was hitherto unknown. This was the first time that it was shown

(a)   (b)

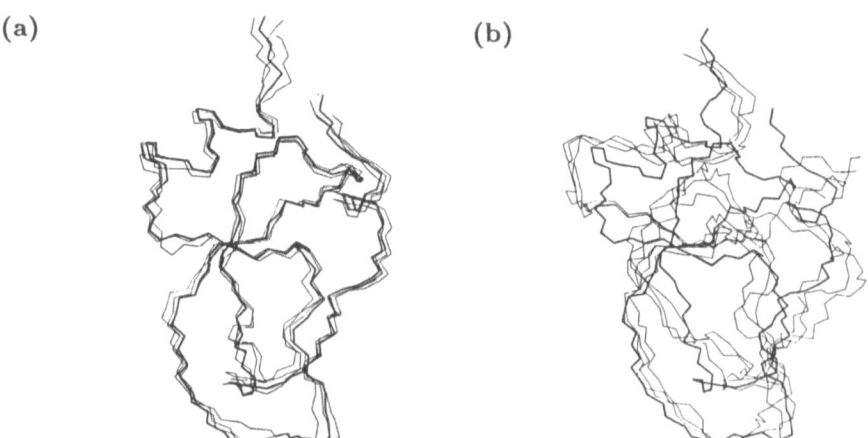

Fig. 3.34a,b. Structure determination from NOE data. The structures were calculated from synthetic NOEs with the distance algorithm (only the main chain is depicted). The thick line represents the X-ray structure that is superposed by structures calculated from 356 (a) and 170 (b) approximate interatomic distances (Havel and Wüthrich, 1985, with permission)

that NMR and X-ray structure analysis can lead to similar results even if the two groups do not know the results of each other's methods in advance.

The number of free variables can in principle be reduced by assuming a standard geometry and accepting as variables only those dihedral angles which are consistent with a simple model construction. Because of the smaller number of degrees of freedom this method can more easily be applied with larger proteins. Its effectiveness was also shown with BPTI (Braun and Gō, 1985).

A method that in principle makes better use of the information available, independent of the actual data, is restrained molecular dynamics. Beginning with as good as possible a starting structure, with this method Newton's classical equations of motion for the N individual atoms i in the macromolecule are in principle numerically integrated. The acceleration d^2r_i/dt^2 of the ith atom with the mass m_i is given by the spatial derivative of the potential V:

$$m_i \frac{d^2 r_i}{dt^2} = -\text{grad} V(r_i, \ldots, r_N) \quad .$$ (3.20)

If the coordinates r_i of all the atoms and their velocities v_i at a time t_0 are known, then the further behavior of all atoms in the macromolecule can be calculated for any time t by solving (3.20) if the potential energy V is known. Equation (3.20) can then be solved by replacing the derivatives dt by the finite time Δt and by calculating all quantities at time $t + \Delta t$ using a linear approximation. An approximation for the velocity v_i at time $t_0 + \Delta t$ is for instance

$$v_i(t = t_0 + \Delta t) = \frac{d^2 r_i}{dt^2}(t = t_0)\Delta t + v_i(t = t_0) \quad .$$ (3.21)

Here, the empirical potential V is a combination of the energy contributions corresponding to the covalent bonds together with other contributions, such as the electrostatic interaction which has in principle to be calculated for all pairs of atoms in the molecule. It was found that molecular dynamics calculations for macromolecules with known crystal structures produce structures that seem plausible, but which in general do not agree with the native structure. This is in fact not surprising because these calculations are performed with considerable approximations. Two of these approximations are quite obvious: the majority of the calculations neglect the interaction with the natural surroundings, the solvent, and calculate in vacuum, although it is well-known from biochemistry that the conformations of proteins depend strongly on the solvent. The essence of the second approximation is that one describes an event, such as the protein folding, which can last several seconds, by a calculation that takes the protein folding process into account for a small fraction of this time, that is for a few picoseconds. That is because even with the large computers available at present it is not possible to take much longer time ranges into account, simply because the individual time steps Δt that must be chosen when integrating the equations of motion are extremely small, typically several femtoseconds. However, this does not greatly matter in the structure determination from NMR data because additional information is introduced into the system. The interatomic distances obtained from the NOEs are introduced as pseudo potentials into (3.20) and then direct the folding toward the fulfillment of the experimental constraints. The typical form of such a pseudo potential is

$$V_{\text{NOE}} = \begin{cases} C_1(r_{ij} - r_{ij}^0)^2 & \text{for} \quad r_{ij} \geq r_{ij}^0 \\ C_2(r_{ij} - r_{ij}^0)^2 & \text{for} \quad r_{ij} < r_{ij}^0 \end{cases} . \tag{3.22}$$

Here, C_1 and C_2 are suitably chosen constants, r_{ij}^0 the distance between the atoms i and j determined from the NOEs and r_{ij} the actual distance between these two atoms. Similar pseudo potentials can, of course, be constructed for other quantities, such as the dihedral angles, and included into the calculation. The development of methods for calculating the 3D structure of proteins is at present still not complete.

The best method for structure determination will use as much additional information about the system as possible in addition to the NOEs. The combination of heterogeneous information can probably best be achieved by using methods of artificial intelligence. However, they will only be superior compared to conventional methods if well-founded physical knowledge implicit in the calculation of molecular dynamics can be integrated into the process.

3.1.7 Solid State NMR of Proteins

Most biological NMR experiments are typical examples of NMR in liquids where a major part of all interactions which depend on the orientation with respect to the external magnetic field are averaged out by fast rotational diffusion of the molecule in solution and only remain perceptible as sources of relaxation. The advantage of NMR in liquids is obvious: in many cases the aqueous solution is the physiological environment in which the molecule is found. Furthermore, spectra in solution are characterized by comparatively narrow resonance lines and therefore can often be more easily interpreted on the atomic level.

However, there are also good reasons for the application of solid state NMR in biology: the orientation-dependent terms that are averaged out in solution contain important, and frequently unambiguously interpretable, additional structural information. Moreover, biological systems also contain components that are not soluble in water, usually forming supramolecular structures. These include membranes, which will be further discussed below, fibrous proteins which build up the cytoskeleton such as collagen, or large systems, which can in fact be composed of individual water soluble components, such as the actomyosin system of the muscle cell or a complete virus particle. Sometimes these systems can be crystallized, and if so, they can of course be analysed by X-ray diffraction methods. In many cases they can be oriented in high static magnetic fields because of their magnetic dipole moments, a property that considerably simplifies NMR spectroscopy.

While for the study of liquids with narrow resonance lines of a few Hz width ^{1}H-NMR is predominant for biological applications, in solid state resonance the situation is different. The large number of protons occurring in macromolecules in combination with the large linewidth in the solid state make separating and assigning of the individual resonance lines much more difficult. Hence, it is in general more favorable to choose a nucleus with low natural abundance, such as ^{2}H, ^{13}C or ^{15}N, and to incorporate it selectively into one or a small number of positions in the molecule. In this way the problems of assignment and superposition are elegantly solved at the same time.

In the case of small molecules in single crystals one can also, in proton NMR, obtain comparatively well-resolved spectra with highly sophisticated NMR methods. Two important methods of line narrowing are multipulse NMR and magic angle spinning (MAS). These methods provide an equivalent of averaging Brownian molecular motion in liquids (Fig. 3.35) by fast spin flips or by very

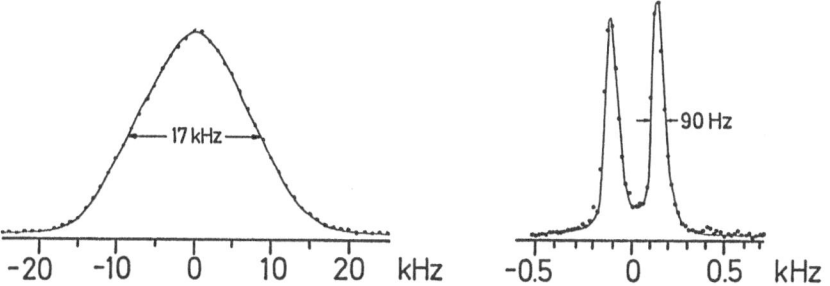

Fig. 3.35. Line broadening in solid state NMR. 90 MHz ^{1}H-NMR spectra of a ferrocene single crystal. (*Left*) normal solid state spectrum, (*right*) WAHUHA multipulse spectrum (Waugh et al., 1968)

fast rotation (approx. 10 kHz) around an axis that forms a 'magic angle' θ of about 54.7° ($3\cos^2\theta - 1 = 0$) with respect to the external magnetic field.

What sort of interactions are especially suitable for structure determination of biological macromolecules in the solid state? The dipolar interaction was already used for measuring the distance between nuclei that are not directly coupled. In the solid state one does not determine the NOE, which depends on the mutual dipolar relaxation, but rather the line splitting $\Delta\nu$ by dipolar interaction, which can be directly measured. For two non-equivalent spin $\frac{1}{2}$ particles A and B at a distance r_{AB} it is given by

$$\Delta\nu = \frac{\mu_0 \hbar \gamma_A \gamma_B}{8\pi^2} \frac{3\cos^2\theta - 1}{r_{AB}^3} . \qquad (3.23)$$

The constants $\hbar, \mu_0$ and γ have their usual meaning and θ is the angle between the vector connecting both nuclei and the direction of the external magnetic field B_0. In a statistically oriented sample θ can assume all possible values. For this reason one obtains a spectrum that consists of a superposition of the spectra of all possible orientations, a powder spectrum. In an oriented sample, on the other hand, θ always has the same value and one obtains a simple doublet spectrum. Since in general the distance between the atoms is also known if they are connected by a covalent bond, there exists a simple relation between the line splitting and the orientation of the covalent bond in space. For the protein structure the orientation of the main chain is of particular importance. Possible dipolar interactions that can be determined are, for instance, the ^{15}N-^1H coupling and the ^{15}N-^{13}C coupling of the peptide bond, and the ^{13}C-^1H coupling of the $C\alpha$-group.

Another orientation-dependent interaction is the anisotropy of the chemical shift that can be important with ^{15}N and ^{13}C. The measured chemical shifts depend on the orientation of the main axes of the corresponding tensor with respect to the magnetic field. Here one is dealing with a number of parameters that must be known, but that can be taken from model substances at least to a first approximation.

The same holds for another tensorial quantity, the quadrupole coupling, which can in principle again be used to determine the orientation of bonds.

These methods can successfully be applied to determine the 3D structure of proteins, as was shown for the coat protein of the fd-virus (Opella et al., 1987). However, at the moment it is difficult to foresee what practical importance solid state NMR will have in this field in future because the experimental efforts needed, in particular the necessary isotope labelling, are rather extensive.

3.2 NMR Spectroscopy of Nucleic Acids, Polysaccharides and Lipids

Besides proteins there are other biological macromolecules that are no less important for the maintenance of biological functions. Since most NMR methods used for investigating proteins can be directly transferred to other macromolecules, we shall restrict ourselves in this section to the characteristic features of NMR studies of nucleic acids, polysaccharides and lipids.

3.2.1 Composition and Structure of Nucleic Acids

Nucleic acids hold a key position in storing and expressing hereditary information. Like proteins they are linear macromolecules that are composed of a varying combination of a small number of different subunits, the nucleotides.

NMR of nucleic acids follows in principle the same rules as NMR in proteins. However, the determination of the spatial structure is frequently simpler because the rules of base pairing considerably reduce the possible number of spatial structures.

The basic units of deoxyribonucleic acids (DNA) are the deoxyribonucleotides, and those of the ribonucleic acids the ribonucleotides (Fig. 3.36). In analogy to the amino acids in proteins these subunits differ only in their side chains, which for DNA consist essentially of the pyrimidine derivatives cytosine and thymine and the purine derivatives adenine and guanine. RNA is built up from the same side chains with the exception that the base thymine is replaced by uracil (Fig. 3.37). In addition to these main basic components of the nucleic acids a smaller number of derivatives of these bases, the so-called rare bases, can be found.

DNA exists usually as a double helix composed of two complementary strands which can occur in several different forms; their importance for regulating and controlling transcription is under intense discussion at present (Fig. 3.38). Normally RNA occurs in the monomer form. Its most important forms are messenger

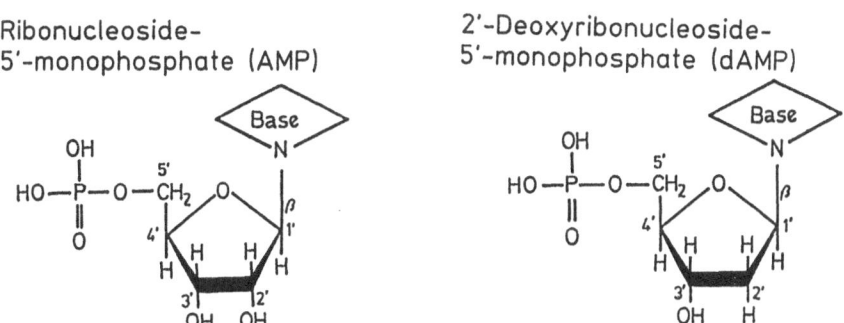

Fig. 3.36. Composition of ribonucleotides and deoxyribonucleotides

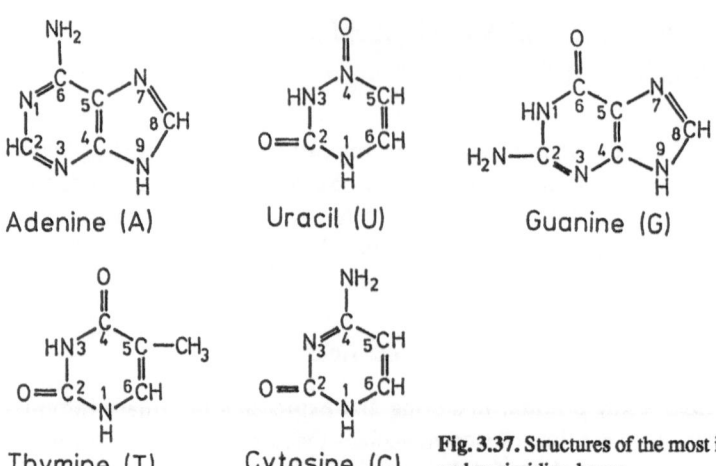

Adenine (A)

Uracil (U)

Guanine (G)

Thymine (T)

Cytosine (C)

Fig. 3.37. Structures of the most important purine and pyrimidine bases

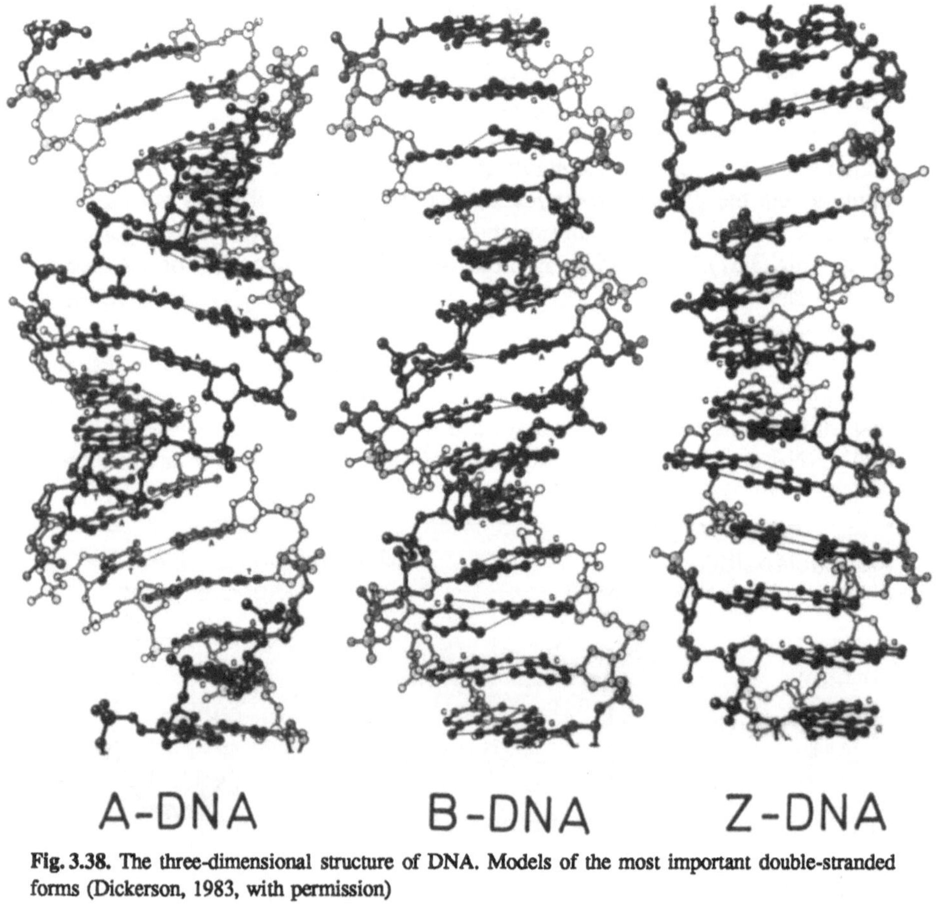

A-DNA   B-DNA   Z-DNA

Fig. 3.38. The three-dimensional structure of DNA. Models of the most important double-stranded forms (Dickerson, 1983, with permission)

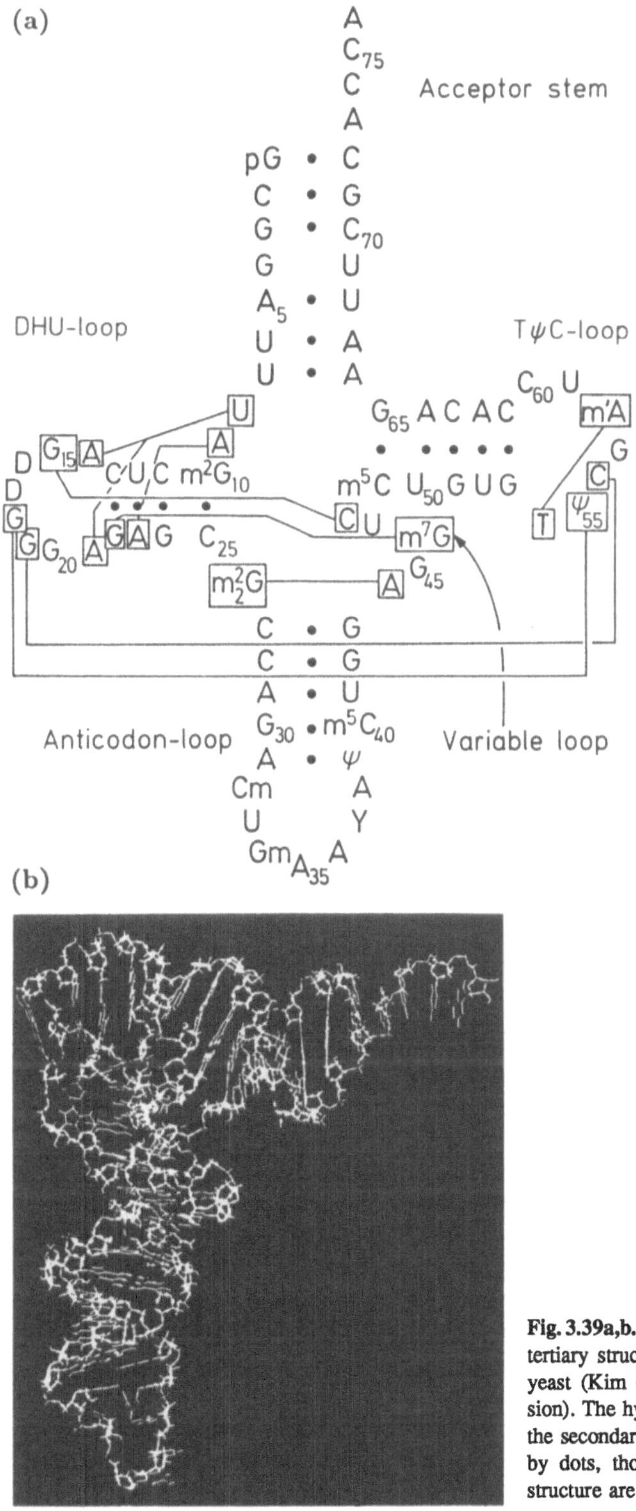

(a)

Acceptor stem

```
              A
              C₇₅
              C
              A
        pG •  C
        C  •  G
        G  •  C₇₀
        G     U
        A₅ •  U
        U  •  A
        U  •  A
```

DHU-loop

TψC-loop

```
                       C₆₀ U
              G₆₅ A C A C        m'A
               • • • • •              G
   C'U C m²G₁₀    m⁵C U₅₀ G U G     C
G₁₅ A    A                      T  ψ₅₅
 • • • •          C U     m⁷G
 A GA G  C₂₅                    G₄₅
G₂₀       m²₂G        A
```

Anticodon-loop

Variable loop

```
        C • G
        C • G
        A • U
   G₃₀ • m⁵C₄₀
        A • ψ
       Cm    A
       U     Y
      Gm A   A
         ₃₅
```

(b)

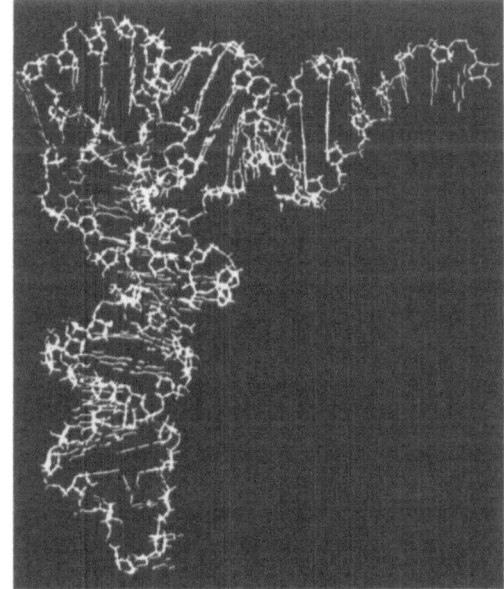

Fig. 3.39a,b. Secondary structure (a) and tertiary structure (b) of rRNA^{Phe} from yeast (Kim et al., 1974, with permission). The hydrogen bridges stabilizing the secondary structure are symbolized by dots, those stabilizing the tertiary structure are represented as solid lines

RNA (mRNA), transfer RNA (tRNA) (Fig. 3.39) and ribosomal RNA (rRNA). Both DNA and RNA undergo biologically important interactions with proteins; the numerous interactions of DNA with proteins are either directly involved with reading the hereditary information or with its regulation. In eucaryotes DNA is in addition permanently associated with proteins, the histones, that participate directly in the building up of the chromosomes. Similar interactions are also exhibited by RNA that as rRNA is rigidly associated with the proteins of which the ribosome is composed. During the transcription and translation mRNA and tRNA interact in various ways with the enzymes that participate in these processes. It is therefore clear that investigating the protein-nucleic acid interaction is one of the most interesting fields to which biological NMR can contribute.

3.2.2 NMR Investigations of Nucleic Acids and Nucleic Acid Protein Complexes

The NMR of nucleic acids has some special features as compared to the NMR of proteins. Since the bases of the nucleotides are aromatic systems, one finds comparatively large conformation dependent chemical shifts that lead to strongly structured NMR spectra. The main chain of the nucleic acid chain contains phosphate diesters; ^{31}P-NMR has proved to be sufficiently sensitive and thus a suitable method for investigation in addition to ^{1}H-NMR. Mg^{2+} ions bind to tRNA and stabilize the structure. Figure 3.40 shows phosphorus resonance spectra of tRNA^{Phe} with different Mg^{2+} concentrations. The phosphorus spectrum exhibits a number of well-resolved single resonances which change their resonance position depending on the concentration. With ^{1}H-NMR all imino protons that participate in the base pairing can be resolved and assigned even in the 1D spectrum (Fig. 3.41). These are particularly important for the structure determination because the structure of nucleic acids is to a large extent determined by the base pairing.

The procedure for sequential assignment of the resonance lines and the calculation of the 3D structure is the same as with proteins. In principle, the nucleic acids can adopt many different structures. The main chain of the subunits has a total of six torsional angles; one of these, however, is situated in the ribose ring and thus possesses only a restricted range of rotation, which depends on the size of the four other endocyclic torsional angles. The position of the base relative to the ribose is determined by one single torsional angle in the glycosidic bond. Within the ribose itself all protons are scalar-coupled; this means in principle that the resonances belonging to a ribose residue can be found with a COSY-type experiment. To the base itself there is only a small J-coupling which cannot always be detected with higher molecular weights. Within the base the J-coupling provides only incomplete information and the nuclear Overhauser effect must be used in addition. In distinction from the proteins, the individual subunits are again scalar-coupled via the phosphate groups, however, the heteronuclear ^{1}H-^{31}P coupling must now be used for the sequential assignment (Fig. 3.42).

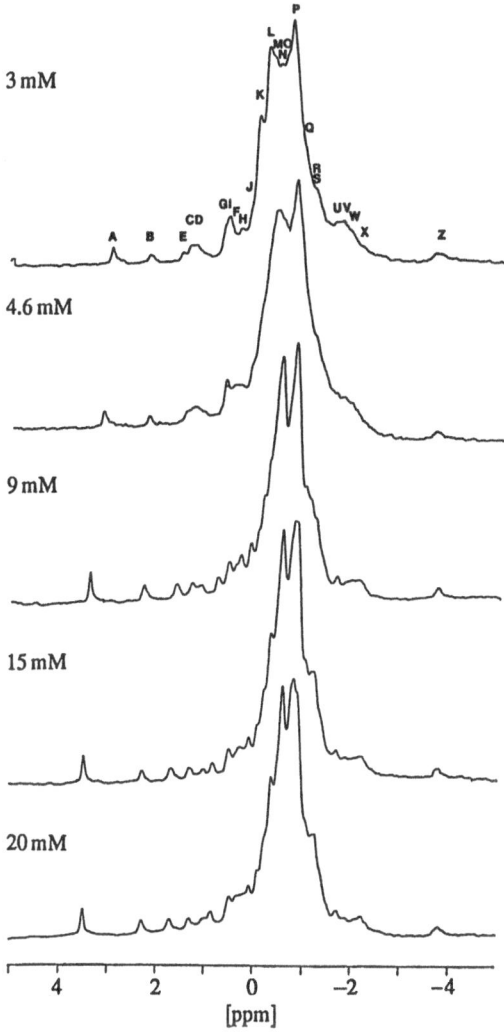

Fig. 3.40. Interaction of Mg^{2+} ions with tRNA^{Phe} molecules. 202 MHz ^{31}P-NMR spectra of tRNA^{Phe} from E. coli at various Mg^{2+} concentrations. The tRNA concentration is 0.34 mM (Hyde and Reid, 1985, with permission)

Starting from a phosphorus resonance of a phosphate group one finds a coupling to the 3'-proton, to the 5'-CH$_2$ group and a still sufficient coupling to the 4'-proton of the deoxyribose. This means that starting from the 3'-protons one can proceed to the corresponding phosphate resonance and from there directly to the 4'-proton resonance of the next residue. The correct assignment can additionally be checked by including the 5'-proton resonance. Finally, the ^{1}H-^{31}P-^{1}H RCT spectrum provides direct information on the 3'- and 4'-proton resonances that belong to two neighboring deoxyribose rings.

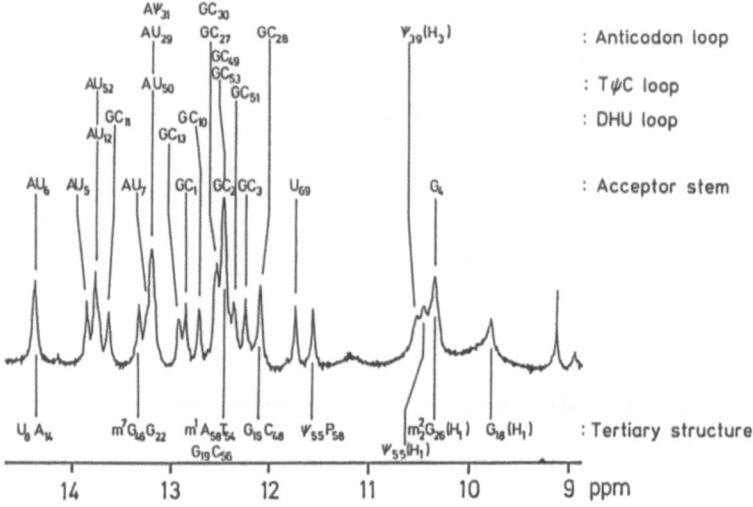

Fig. 3.41. 500 MHz ^{1}H-NMR spectrum of the imino protons of tRNA^{Phe} from yeast (Heerschap et al., 1983, with permission)

The information obtained from the scalar coupling and the NOE is partly redundant so that the spin systems of a nucleotide can be interpreted with comparative certainty. Besides depending on the spatial structure there are sequential NOEs, which can be used to identify neighboring nucleotides in the sequence, and strong NOEs between the strands of the base paired structures.

Calculating the structure of DNA double helices is comparatively simple since one usually has a comparatively good starting model because of the base pairing and the regularity of the structures (A-, B-, Z-DNA); after having selected the best regular structure one has to refine the results. This is similarly true for the single strand DNA or the single strand RNA if they possess a defined structure, such as for instance the tRNA.

If the structure of a DNA component is known and its resonance lines are assigned, its interaction with specific proteins can be studied. An example is the interaction with the lac-repressor protein that hinders the transcription of lactose-specific enzymes if the cell does not contain lactose. The NMR structure of the lac-repressor head part that is responsible for the binding to the DNA was published some time ago (Kaptein et al., 1985). The relative position of the protein with respect to the DNA can be determined from the NOESY spectrum of the complex if a few NOEs are known. Experimentally, using ^{1}H-NMR NOEs between three different aromatic rings of the lac-repressor (Tyr-7, Tyr-17 and His 29), certain nucleotides of the lac-operator were found. With these estimates of the distances one obtains the structure of the complex in which the lac-repressor is turned by 180° compared to the predictions from pure model building attempts (Fig. 3.43).

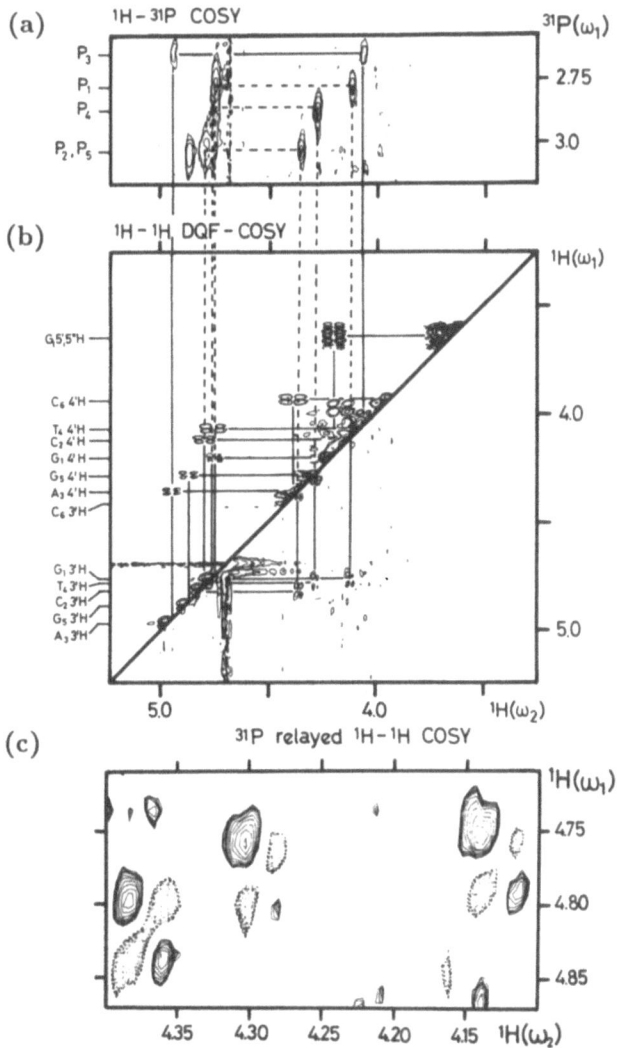

Fig. 3.42a–c. Sequence specific assignments in a fragment of DNA. (a) ^{31}P-^{1}H heteronuclear correlated 2D NMR spectrum of d(GCATGC)$_2$. Proton resonance detection at 360 MHz. (b) Corresponding double quantum filtered homonuclear ^{1}H-COSY spectrum. (c) ^{31}P relayed ^{1}H-^{1}H-RCT spectrum (Frey et al., 1985, with permission)

3.2.3 Composition and Structure of Polysaccharides

The third group of biological macromolecules that are built up from simple subunits are the oligo- and polysaccharides. They are composed of simple monosaccharide units that are closely bound to each other. The polysaccharides have various biological functions; they serve as reserve substances, such as starch,

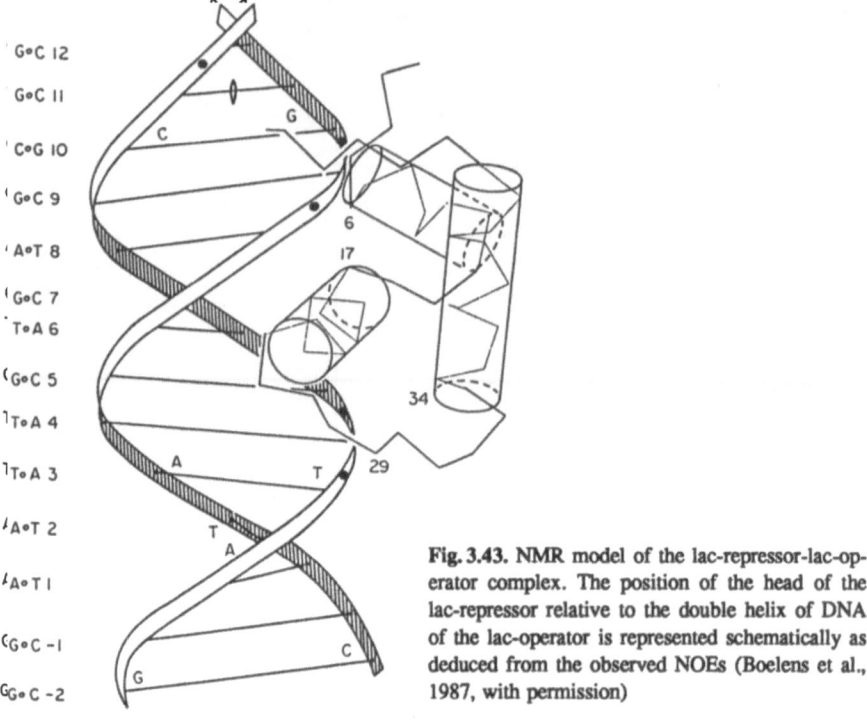

Fig. 3.43. NMR model of the lac-repressor-lac-operator complex. The position of the head of the lac-repressor relative to the double helix of DNA of the lac-operator is represented schematically as deduced from the observed NOEs (Boelens et al., 1987, with permission)

and as structure elements, such as cellulose. An important function is that of surface elements which are used for cell-cell recognition and as receptors. These are mostly oligosaccharides or small polysaccharides bound to lipids, and their primary structure determines their properties.

The basic components of the polysaccharides, the monosaccharides, show a larger variety than the amino acids. However, monosaccharides with five carbon atoms, pentose, and with six carbon atoms, hexose, are much more frequently found than other carbohydrates. The monosaccharide subunits are usually bound to each other via glycosidic bonds (Fig. 3.44). In contrast to proteins, one finds mostly branched chains because with saccharides each hydroxyl group can form a branching point.

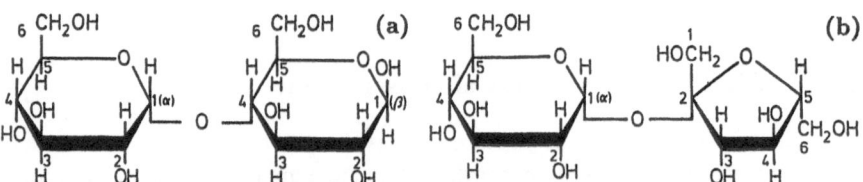

Fig. 3.44a,b. Structures of polysaccharides. The disaccharides maltose (a) consisting of two molecules of glucose and saccharose (b) consisting of glucose and fructose are shown

3.2.4 Structure Determination of Polysaccharides

The chemical analysis of polysaccharides is quite difficult; for this reason the subject of many investigations of saccharides is the determination of their chemical structure. The ring of the saccharide molecule itself can adopt different conformations: frequently occurring are the chair and boat conformations. Both are characterized by typical J-coupling constants and can be distinguished quite easily with proton resonance. Also in both conformations the distances between the individual ring protons differ characteristically; hence, one expects in principle different nuclear Overhauser effects. In contrast to the routine sequencing of proteins, no such simple sequencing method exists in the chemistry of saccharides. For a complete description of the structure it is not only necessary to know the binding patterns of the monosaccharides, but also the type of binding, which means that in addition one must determine whether it is an α- or β-glycosidic bond (Fig. 3.44). For these purposes 2D NMR proved to be very suitable. If

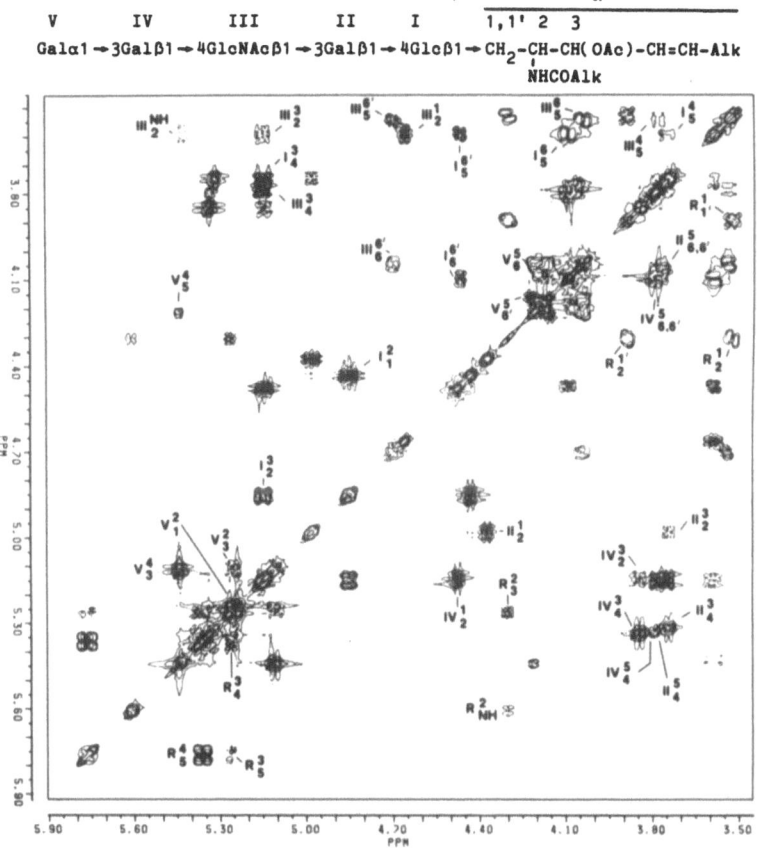

Fig. 3.45. 500 MHz ^{1}H-NMR COSY spectrum of a peracetylated glycosphingolipid that was isolated from membranes of erythrocytes (Dabrowsky et al., 1986, with permission)

only the binding pattern of the monosaccharides is of interest, the assignment of the NMR lines can be considerably simplified by forming suitable derivatives (Fig. 3.45).

Of course, the 3D structure of polysaccharides is of biological interest, too, which again is determined by the rotation around the single bonds in the non-cyclic parts of the molecules and by the conformation of the ring systems. These conformations can again be determined by measuring coupling constants and nuclear Overhauser effects. However, in this case it is rather difficult to assess the biological relevance of the structures found. The binding of polysaccharides to lipids and proteins and their embedding into the cell membrane have probably a strong influence on their special structures so that measurements of isolated polysaccharides even if they are performed in aqueous solution are not very representative for the intact structure. On the other hand, such measurements probably still provide the best possible information that can be obtained on the structures of polysaccharides.

3.2.5 Investigations of Biological Membranes

The lipids form a large heterogeneous class of organic molecules that can be extracted with non-polar solvents. Besides isolated molecules, such as fat-soluble vitamins and steroid hormones, where NMR can contribute to determining the chemical structure, those phospholipids are of particular importance that form planar, supramolecular structures, the lipid double membranes. On account of their importance, membranes have so far been investigated by all biophysical methods available, including NMR.

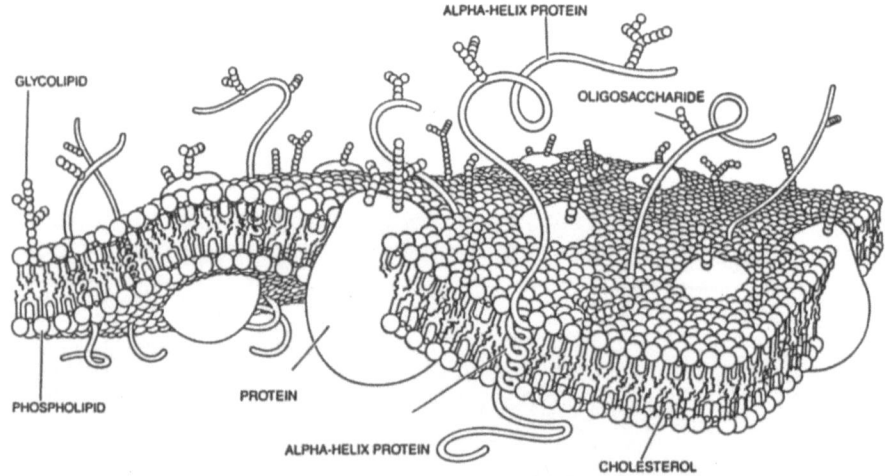

Fig. 3.46. Model of a plasma membrane. In biological membranes which are essentially composed of phospholipids and cholesterol the embedded proteins and glycolipids can freely diffuse in the plane. The oligosaccharides' residues are mainly found at the outer side (*above*) of the membrane (Bretcher, 1985, with permission)

When investigating membranes one always deals with a large number of molecules. Natural membranes form a very complex system that consists of many different lipids and proteins embedded into the lipid membrane (Fig. 3.46). Hence, it is extremely difficult to obtain such detailed information on individual molecules as in the case of soluble proteins. One can hope to find reliable answers to fundamental questions, like organization and state of motion, only in well-defined model systems, such as micelles or liposomes of defined composition. These results can then later be transferred to natural membranes.

Depending on temperature and composition, membranes exist in different physical phases. At low temperatures they behave like solids, at higher temperatures they are in a liquid crystalline state with high mobility in the plane of the membrane. In the liquid crystalline state one finds that lateral diffusion constants are almost as high as those in water. Biologically active membranes under physiological conditions are usually in the latter state. The mobility restricted to one plane leads to comparatively broad lines in the NMR spectrum because an isotropic motion is not possible. Below the transition temperature extremely broad lines of many ppm linewidth are found as they are typical for solid state resonances.

Since phospholipids contain a phosphate group, liposomes containing phospholipids can be observed by using phosphorus NMR. Above the phase transition temperature the different phospholipids in artificial membrane vesicles can be separated under favorable conditions because of their chemical shifts (Fig. 3.47). In small vesicles even the chemical shifts of the lipids on the inner and on the outer sides differ by several Hertz so that they can be observed separately as

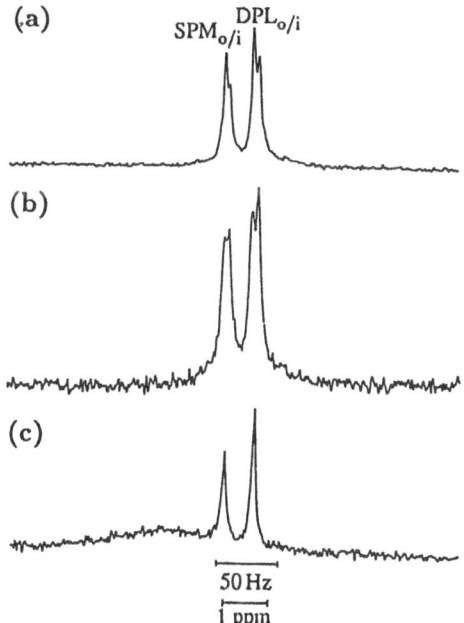

Fig. 3.47a–c. ^{31}P-NMR spectrum of simple artificial membranes. 36.4 MHz phosphorus resonance spectra of vesicles of membranes consisting of a mixture of sphingomyelin and dipalmitoyllecithin. Temperature 50°. (a) Vesicle in Tris-buffer, (b) after adding 5 mM $CoCl_2$, (c) after adding 7 mM $CoCl_2$. $DPL_{o/i}$, dipalmitoyllecithin (outside/inside), $SPM_{o/i}$, sphingomyelin (outside/inside) (Berden et al., 1975, with permission)

well. To facilitate this, suitable paramagnetic substances can be added either on the inner or the outer sides for which the permeability through the membrane is low. The resonance lines of the lipids that are in contact with the paramagnetic substances are then strongly broadened and are no longer visible. Phosphorus resonance of liposomes is also an example of the fact that a higher magnetic field does not always lead to a higher resolution, because with phosphorus the relaxation due to the anisotropy of the chemical shift contributes considerably to the total relaxation. Since this effect is proportional to the square of the field strength, the line broadening can overcompensate the resolution which varies only linearly with the magnetic field. In the system mentioned above the resonances of the inner and outer parts can indeed no longer be distinguished at higher fields (Fig. 3.48). Of course, membrane lipids can also be investigated with ^{1}H-NMR. Here the signals of different groups in the lipid can be observed separately. ^{13}C and ^{1}H-NMR permit one to measure lipids that are selectively labelled in certain positions and thus to selectively observe even atoms in the long hydrocarbon chains.

One property of lipid membranes is the order of the lipid molecules. In an extreme case all molecules are well ordered as in the crystalline state. The other limit is the statistical order (disorder) that is more or less extensively found in liquids and powders. For a quantitative description so-called order parameters are introduced. They measure how certain quantities (for example, the direction of a vector that connects certain atoms in the molecule) vary over the ensemble.

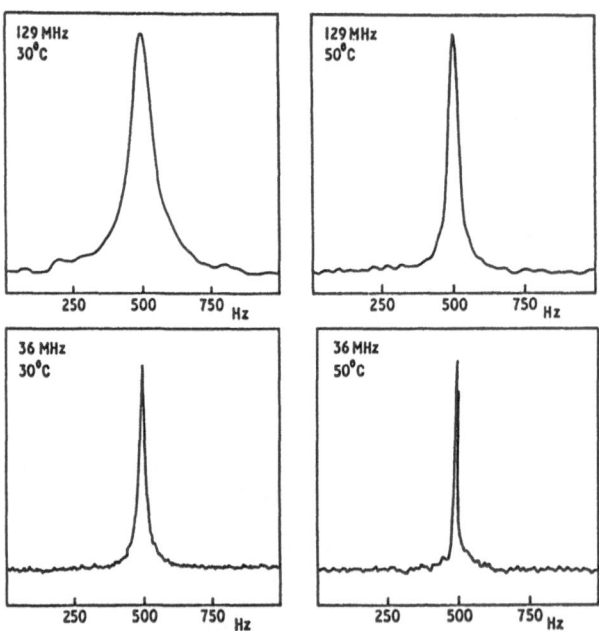

Fig. 3.48. ^{31}P-NMR spectra of artificial membranes (vesicles of dipalmitoyllecithin) measured at various temperatures and magnetic fields (Berden et al., 1974, with permission)

A simple form of an order parameter would be one that adopts the value 1 if the corresponding quantity is exactly identical for all molecules and becomes zero if it takes all possible values corresponding to a statistical distribution. In practice, however, one often defines other order parameters which cannot be understood that easily, but which are directly connected with spectroscopically measurable quantities (see below).

Relevant insight into the organization of membranes was provided by the deuterium resonance of oriented membranes. The latter can be produced experimentally by placing a single bilayer between two thin glass plates which determine the orientation of the surface of the membrane. In order to obtain a better signal-to-noise ratio, a whole pile of these glass plates can be used as the NMR sample without problems. In orienting the membranes one passes from the area of pure resonance in liquids, where the majority of the interactions are averaged out by the isotropic motion, towards the situation of resonance in the solid state. In oriented systems the quadrupole coupling leads to a splitting of the resonance lines $\Delta\nu$, which depends on the orientation of the system with respect to the external magnetic field. In the liquid crystalline state the rotational diffusion around an axis perpendicular to the surface of the membrane is usually fast enough to average out differences within this plane. The quadrupole splitting $\Delta\nu$ depends now on the angle θ between the C-D bond direction and the normal to this plane, and on the angle that the normal forms with the direction of the external magnetic field. The corresponding order parameter S_{CD} is then defined as

$$S_{CD} = \tfrac{1}{2}\langle 3\cos^2\theta - 1\rangle \quad . \tag{3.24}$$

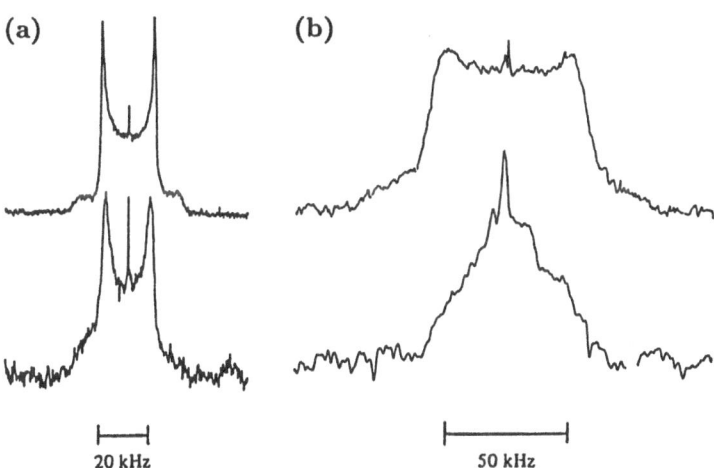

Fig. 3.49a,b. 46.1 MHz deuterium resonance spectra of reconstituted sarcoplasmic vesicles. The naturally occurring lipids were replaced by 1,2-dielaidoyl-sn-glycerol-3-phosphocholine (DEPC) labelled in positions 9 and 10 by deuterium. (*Top*) Membrane without protein, (*bottom*) membrane containing the natural membrane proteins. Measurements at 25° C (a) and 4° C (b) (Seelig and Seelig, 1980, with permission)

In the case of liposomes rather than oriented membranes one averages over all possible directions that the normal takes with respect to the surface of the membrane, one finds a maximum for the perpendicular orientation. Hence, the doublet signals of these molecules predominate in the total spectrum and give two maxima. The corresponding splitting is

$$\Delta \nu = \frac{3}{4} \frac{e^2 q Q}{h} S_{CD} \quad , \qquad (3.25)$$

where h is Planck's constant, eQ the quadrupole moment of the deuterium and eq the electrical field gradient along the C-D bond. Figure 3.49 shows the deuterium resonance spectra of reconstructed sarcoplasmatic vesicles below and above the phase transition temperature. The fatty acids of the membrane lipids are specifically labelled in positions 9 and 10. By embedding the sarcoplasmatic ATPase in the membrane the quadrupole splitting decreases, which means the order parameter decreases as well. Below the transition temperature one finds the broad spectrum typical of the gel state. Here the embedding of the protein in the membrane has an even more drastic effect on the NMR spectrum.

4. NMR Tomography

In this chapter we shall present another aspect of NMR in biomedicine, namely magnetic resonance imaging (MRI). Other terms for this spatially resolved NMR are NMR (or MR) tomography and NMR zeugmatography. The quantity measured for imaging by NMR is the nuclear magnetization in a given volume element. In this respect it differs fundamentally from the established method of X-ray computer tomography, which measures the X-ray absorptions of the various tissues, bones, etc.

As we have already seen, the magnetization detected in an NMR experiment depends not only on the nuclear spin density but also on additional parameters. The most important ones for imaging are the longitudinal and the transverse relaxation times, which are influenced, as is the spin density, by the structure and composition of the tissue. Hence, in distinction to X-ray computer tomography, there are at least three parameters which can be used for describing different tissues. Thus it is possible to obtain contrast by applying suitable pulse sequences even in tissues that are homogeneous with respect to one parameter, simply by emphasizing some others. In addition to the advantage that no ionizing radiation is required, this contrast behavior is the reason why in many cases NMR tomography is superior to X-ray computer tomography, although the signal-to-noise ratio of the latter can be in principle much higher.

4.1 Basic Principles of Imaging

For determining the structure of biomolecules with high resolution NMR the signal must be as similar as possible for all volume elements of the sample volume, and therefore a static magnetic field B_0 of maximum possible homogeneity is required. In contrast to this, with space selective NMR one needs to obtain different NMR signals from the various volume elements of the heterogeneous sample, for instance the human body. This is in principle achieved by superposing magnetic field gradients on the static magnetic field B_0, so that particular nuclear spins, selected because of their abundance and their large magnetic moment, usually the nuclei of hydrogen atoms (protons), are in resonance in different parts of the sample at different frequencies.

The usual procedure with spatially resolved NMR is first to choose a planar thin slice by applying a magnetic field gradient. The principle is explained

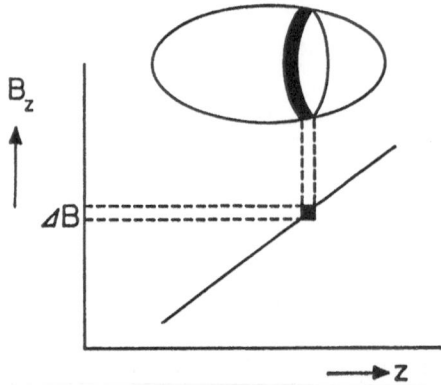

Fig. 4.1. Principle of selective excitation. Owing to the superposition of the static magnetic field B_0 and a constant field gradient G_z only nuclear spins within a thin slice are excited by a frequency selective radio frequency field

in Fig. 4.1. With the method of selective excitation a constant field gradient is applied along a certain direction, so that the size of the static magnetic field increases or decreases linearly along this gradient, while the direction of the magnetic field remains unchanged.[1] In Fig. 4.1 we have chosen the z direction as that of the gradient, but in principle we could have defined any direction in space. If the sample is irradiated with a radio frequency within a narrow frequency range, the resonance condition (1.3) is satisfied only for the nuclear spins within the thin slice, and hence only these nuclear spins can contribute to the signal.

In order to obtain a 2D picture of this slice, we divide it into a matrix of $n \times n$ picture elements (pixels) $A_{p,q}$, where n depends on the desired resolution consistent with the achievable signal-to-noise ratio, for instance typically $n = 256$. Now we have to determine the NMR signals of these pixels $A_{p,q}$ and then to show them as a picture, for example on a video screen. Here there are in principle several possibilities. The simplest one is to isolate all n^2 pixels one after the other; this sequential point method obviously takes too much time and is therefore of no practical importance. In order to save time, the n elements of one row or column, or still better the NMR signals of all n^2 pixels in a slice, are measured simultaneously, a method that is termed 'planar imaging'. In many cases one or several of such slices will suffice. On the other hand, it is also possible to obtain an image of the whole 3D sample by viewing one slice after the other.

Finally it is also possible to measure simultaneously an image of the whole 3D sample instead of restricting oneself to one slice, but this takes much too long with most techniques. For this reason methods for shortening the recording time are here of particular practical importance.

4.1.1 Two-Dimensional Projection-Reconstruction

So far two methods have proved to be suitable for the practical application of planar imaging: the 2D projection-reconstruction and the 2D Fourier-imaging.

---

[1] In the literature one frequently finds the incorrect expression 'linear field gradient': in fact the magnetic field depends linearly on the spatial coordinates, but the gradients are constant.

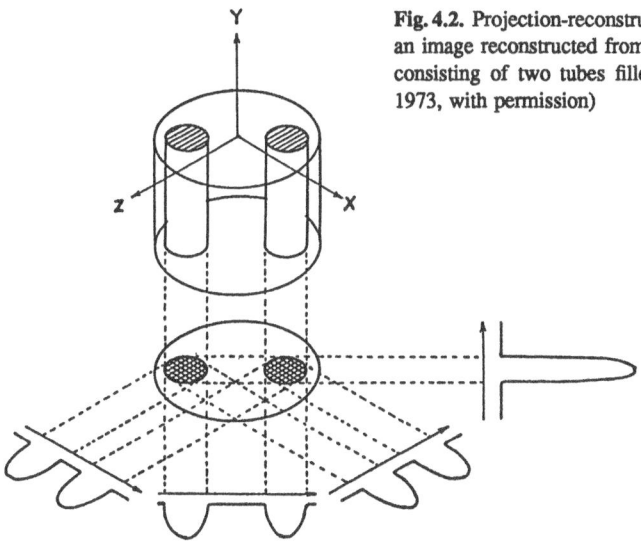

Fig. 4.2. Projection-reconstruction method. Scheme of an image reconstructed from NMR data of a phantom consisting of two tubes filled with water (Lauterbur, 1973, with permission)

The 2D projection-reconstruction method, with which the first published image was produced by *Paul Lauterbur* (Lauterbur, 1973), is nowadays only of historical interest. It is carried out in the following way: first, a slice is defined as explained above. Then a field gradient is applied in the plane of this slice and the NMR signal is detected so that one obtains a 1D projection of the proton spin density along this field gradient. The direction of this field gradient is then turned around in the plane successively by the same angle $180°/n$, n times, so that a series of 1D projections in this plane is obtained. By means of a computer an image of the proton spin density in this plane is then reconstructed in a manner similar to X-ray tomography (Fig. 4.2).

4.1.2 Two-Dimensional Fourier Imaging

The methods almost exclusively applied at present are several variations of 2D Fourier imaging, which is closely related to the 2D Fourier transform spectroscopy discussed above. It was initiated by Richard Ernst and his collaborators (Kumar et al., 1975). At first, the nuclear spins in a thin slice perpendicular to the direction of the slice selection gradient are excited or inverted by applying this gradient simoultanously with a radio frequency pulse, for instance a 90° or 180° pulse. The principle of the procedure is illustrated in Fig. 4.3; the direction of the slice selection gradient is again the z direction. After the first excitation pulse in the case shown here a gradient G_x is applied for a time t_x along the x axis and afterwards a gradient G_y for a time t_y along the y axis. The NMR signal is then digitized and recorded for n equidistant values of t_y. This process is repeated n times for n equidistant values of t_x so that a matrix of n^2 measured points is obtained. The two-dimensional picture is then obtained by a two-dimensional Fourier transformation.

Let us analyse this procedure in more detail. Let the proton spin density in the slice at the point (x, y) be $\varrho(x, y)$. Neglecting relaxation effects the NMR signal ds of a pixel $dx\,dy$ at a point x, y (C is a proportionality constant) is

$$ds = C\varrho\,dx\,dy\,e^{i\gamma_I B t} = C\varrho\,dx\,dy\,e^{i\omega t} \quad . \tag{4.1}$$

At time t_x the magnetic field B is

$$B = B_0 + xG_x \quad , \tag{4.2}$$

where B_0 denotes the value of B at (0,0). Correspondingly during the following time t_y is

$$B = B_0 + yG_y \quad . \tag{4.3}$$

After inserting (4.2) and (4.3) into (4.1) the NMR signal ds of a pixel at the frequency $\omega = \gamma_I B_0$ with phase-sensitive detection at frequency ω_I

$$ds = C\varrho\,dx\,dy\,e^{i\gamma_I(xG_x t_x + yG_y t_y)} \tag{4.4}$$

and for the whole slice

$$\begin{aligned} s(t_x, t_y) &= C \iint \varrho(x, y) e^{i\gamma_I(\omega_x t_x + \omega_y t_y)} dx\,dy \\ &= CA \iint \varrho(\omega_x, \omega_y) e^{i(\omega_x t_x + \omega_y t_y)} d\omega_x d\omega_y \end{aligned} \tag{4.5}$$

with

$$\omega_x = \gamma_I x G_x, \quad \omega_y = \gamma_I y G_y, \quad A = (\gamma_I^2 G_x G_y)^{-1} \quad . \tag{4.6}$$

The proton spin density $\varrho(x, y)$, the image of which we want to construct, is hence, because of the reciprocity of Fourier transformations, the 2D Fourier transform of our measured quantity $s(t_x, t_y)$.

As we can see, the equations always contain the product $G_x t_x$. It is therefore possible, instead of the procedure described in Fig. 4.3, to keep t_x constant and to give G_x n different values one after the other, which has practical advantages and is therefore usually done (Edelstein et al., 1980).

So many combinations of gradient and pulse sequences already exist today that we have to restrict ourselves within the scope of this introduction to the description of the principles of some simple examples. The simplest pulse sequence consists of a 90° pulse and subsequent detection of the signal, of course, in combination with suitable gradients. Alternatively, one can also saturate the system by irradiating first with another 90° pulse, a method that supplies T_1-weighted images depending on the time interval τ between the two pulses; this is called the saturation recovery method. If one irradiates first with a 180° pulse instead of a 90° pulse, one obtains the 180°-τ-90° pulse sequence described in Sect. 1.1.4 for measuring the spin-lattice relaxation time T_1, where the time τ can again be varied. Since the magnetization is first flipped into the $-z$ direction by

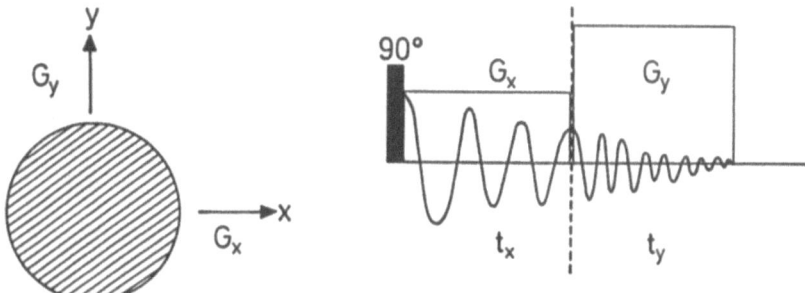

Fig. 4.3. 2D Fourier imaging. After selecting a slice (see Fig. 4.1) gradients G_x and G_y in the x and y direction, respectively, are applied for times t_x and t_y

the 180° pulse, this pulse sequence is called the inversion recovery method. It also produces strongly T_1-weighted images depending on the distance τ between the two pulses.

In the large majority of experiments one observes spin echoes that are either produced by radio frequency pulse sequences or by gradient pulses with changes of signs. The most common radio frequency pulse sequence is the Carr-Purcell-Meiboom-Gill (CPMG) sequence (Fig. 1.8) described in Sect. 1.1.5. In order to improve the signal-to-noise ratio one can, for instance, average over eight echoes after eight 180° pulses (Fig. 4.4). For obtaining different T_2-weighted images, one can always use only a certain echo, for instance the first or the eighth, for imaging.

A CPMG pulse sequence lasts typically 100–300 ms depending on the number of echoes produced. Afterwards, however, one has to wait, for instance 2 s, because of the necessity to complete the longitudinal relaxation process before the next pulse sequence can be started (Fig. 4.4). This means we need approximately 8.5 min for 256 G_x values to record one disk. A much better use of the time can be made if we select and measure additional slices one after the other by shifting the irradiation frequency during this waiting time, as shown in the lower part of Fig. 4.4. This technique is called the multi-echo/multi-slice technique and it is the one generally applied at present. The gain of time with this method is due to the fact that in the time required for recording one slice several slices, for instance eight, can be recorded. However, this does not mean that the recording of a single slice is shortened.

4.1.3 Parameters of the NMR Tomogram

The essential parameters on which the quality and applicability of an NMR tomogram in diagnostics depend are the following:

1. Contrast resolution
2. Spatial resolution
3. Image quality depending on the signal-to-noise ratio

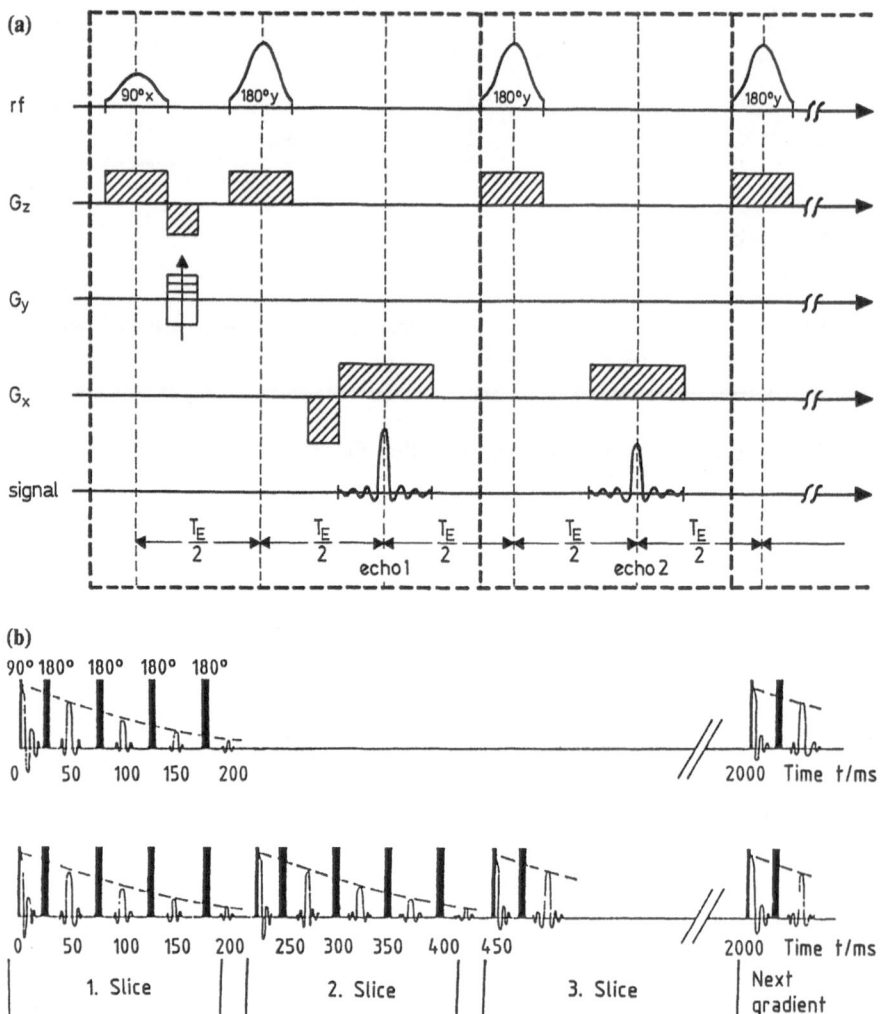

Fig. 4.4a,b. Spin echo techniques. (a) Pulse and gradient sequence for the spin echo technique consisting of a 90° and 180° pulse that can be extended to the multi-echo technique by additional 180° pulses, (b) Extension of the multi-echo techniques (*upper part*) to the multi-echo multi-slice technique (*lower part*)

4. Absence of artifacts
5. Recording time.

Contrast resolution is a particular strength of NMR tomography. Although every tissue is composed of a multitude of different molecules, the observed proton resonance signal originates primarily from the nuclear spins of water in the extracellular and intracellular space. On the one hand, this is due to the fact that water is present in the organism in very high concentration. On the other hand, in the case of very large molecules or molecules immobilized in supramolecular structures the corresponding resonance lines are so strongly broadened that they become to a large extent undetectable. Hence, besides water only small molecules or molecules with high mobility, such as lipids in fatty tissue, contribute to the total signal in proportion to their concentration. Different tissues or those that are changed pathologically have different contents of water, as is shown by the intensity difference in the NMR signal. However, the differences in the relaxation times are much larger because they reflect in a complex manner the interaction of the water molecules with the molecular components of the cells. In general, the signal amplitude is determined by all three factors, the spin density ϱ, and the T_1 and the T_2 relaxation times. With the methods discussed above, the weighting of these factors in the image can be adjusted and thus the image contrast can be varied over a large range.

The other parameters are not independent of each other. If we demand a certain signal-to-noise ratio, there exists, as always with physical experiments, a relation between the precision of the measurement and the time required for performing it. For example, the linear spatial resolution can be doubled by doubling the strength of the gradients. This reduces the signal intensity per pixel and therefore the signal-to-noise ratio by a factor of 4. If one wants to obtain the same signal-to-noise ratio as before, one has to average over 16 recordings, that is the total recording time has increased for a slice of equal thickness by a factor of 16 due to the doubling of the linear resolution. If one wants to double the resolution in the third dimension as well, either with a 3D image or with a 2D image reducing the thickness of the slice by a factor of 2, for the same signal-to-noise ratio a further factor of 4 is required for the recording time, that is, for a doubling of the linear resolution in all three dimensions in space an overall factor of 64 increase in the total recording time is needed.

In practice, for an image of 256×256 pixels (depending on the strength of the gradients) a spatial resolution of about 1 mm can be achieved. For the simultaneous recording of, for instance eight slices, this results in a recording time of the order of 5–10 min. Since such a period is too long for many practical applications, several methods for reducing the recording time have recently been developed, and these will be discussed below. Although these methods have in many cases certainly brought about important progress for the application of NMR tomography in diagnostics, it is quite clear that one cannot break the laws of physics and that each reduction in time represents a compromise in one way or another that has to be paid for by a reduced spatial resolution, by a worse signal-to-noise ratio, by additional artifacts, or by a combination of these.

Increasing the strength of the B_0-field is in principle a possible way to improve the signal-to-noise ratio because the signal is proportional to $B_0^\lambda (1 < \lambda < 2)$. However, in practice there are not only technical and financial limits, but also fundamental physical limits, because with increasing frequency the losses in the body increase and therefore the radiation cannot pass through the body without distortion. This effect has, however, proved to be less disturbing than was originally feared. In general, it can be stated that for in vivo spectroscopy, dealt with in Chap. 5, in which one makes use of the chemical shift that is proportional to B_0, the magnetic field strength should be as high as possible. It is not yet clear which is the optimum field strength for imaging; the magnetic fields used vary in the range between 0.02 and 2 tesla. Below about 0.3 tesla the field for whole body tomographs can be generated with resistive coils, while for higher fields superconducting coils must be used. The latter are more expensive and require liquid helium, but do not dissipate electric power during operation. Besides the lower initial costs, the main advantages of the low field equipment are lower stray fields, lower radio frequency power, no artifacts by chemical shifts and rather short T_1 values because the relaxation time of the protons in the tissue increases with increasing magnetic field. The latter makes higher repetition rates possible, thus compensating partially for the better signal-to-noise ratio at high field by making a shorter measuring time possible. The main advantage of the high field equipment is the better signal-to-noise ratio and therefore a shorter measuring time with equal resolution.

4.1.4 Methods for Reducing the Recording Time

The main disadvantage of NMR tomography is, as was explained above, the relatively long duration of an individual measurement, which is essentially determined by the fact that one must wait several T_1 intervals between two pulse sequences until the magnetization has been reestablished by the spin-lattice relaxation. For a linear resolution of 256 pixels a total of 256 pulse sequences is required, that is, a delay time between successive pulse sequences of 1 second entails 256 seconds, and when averaging over two sets of pulse sequences already $2 \times 256 = 512$ seconds or about 8.5 minutes are needed for recording one slice. Hence, images of the thorax or abdominen (motions of the heart and respiratory system, peristalsis of the intestines) are perturbed by motional artifacts; in the case of periodic motions this can be avoided by a stroboscopic recording. Recently, special gradient pulse sequences (MAST, EXORCIST) have been developed that compensate the perturbing effects of motion, the motion-induced dephasing. However, in any case a reduction of the recording time is desirable for a variety of reasons – we mention here the stress of seriously ill patients, a more efficient use of the equipment, and in the event that a sufficient reduction can be achieved, the possibility of recording time-dependent processes.

The methods for reducing the measuring time as proposed by various groups can in principle be divided into two categories: the fast imaging methods with gradient echoes and those based on a different phase coding of signals (echoes) recorded within one pulse sequence (mostly CPMG sequences) after one single excitation of the spin system. The first method for reducing the measuring time was the echo planar imaging, developed by *Peter Mansfield* (Mansfield, 1977), that combines in a certain sense elements of both methods. A more general application failed at the beginning because it requires very strong gradients and very short switching times. With the progressive development of techniques the situation could change again to make this method become important again in the future.

The technique of gradient echoes has in principle been known for some time. With this technique one obtains, similar to a 180° pulse, a sequence of spin echoes by suitable gradient inversions in corresponding time intervals. However, an essential difference has to be kept in mind: while the use of suitable radio frequency pulse sequences such as CPMG (see Sect. 1.1.5) allows dephasings due to inhomogeneities of the B_0-field to be rephased, thus leading to a spin echo, only dephasing due to gradients can be rephased by gradient inversion. These methods therefore require highly homogeneous B_0-fields; even the inhomogeneity of the magnetic susceptibility within the sample is a problem.

Since 1985 a team working in Göttingen has developed a fast imaging method on the basis of gradient echoes (Haase et al., 1986). The essential idea of this method, called FLASH (Fast Low Angle SHot) by the authors, is the excitation of the spin system with small flip angles of 15° to 30° instead of 90°, which permits repetition times that are considerably shorter than the spin-lattice relaxation time T_1. The reason for this was explained above in Sect. 1.4.2.

To read the image information a gradient echo is used. The details of the FLASH sequence are plotted in Fig. 4.5. A radio frequency pulse of 15° to 30° is irradiated simultaneously with the slice selective gradient; in a second phase a rephasing with respect to the slice selection gradient and a predephasing with respect to the reading gradient occurs. At the same time the phase coding gradient is switched on with n different, linearly equidistant increasing amplitudes. Finally, after changing the sign of the reading gradient the signal is observed in the form of a gradient echo. Afterwards the remaining transverse magnetization is destroyed with a gradient pulse (spoiler gradient). Subsequently, the same scheme can be repeated immediately without time delay. The most favorable values of the flip angle and the repetition time depend on the T_1 times.

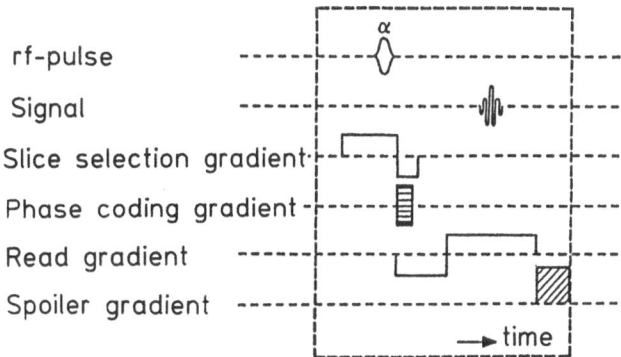

Fig. 4.5. Pulse and gradient sequence with the FLASH method

Conversely, different T_1-weighted images can be produced by suitably choosing the flip angles. Quantitatively, the signal with FLASH is proportional in addition to the equilibrium magnetization M_0 to

$$\frac{(1 - e^{-T_R/T_1}) \sin \alpha}{1 - e^{-T_R/T_1} \cos \alpha} e^{-T_E/T_2^*} , \tag{4.7}$$

where T_R is the repetition time, T_E the echo time, T_2^* the effective transverse relaxation time and α the flip angle of the magnetization. The optimum flip angle α_E for maximum intensity is in the case of FLASH equal to the Ernst angle (1.53). In this way repetition times of 10–20ms can be achieved; with 256^2 pixels this corresponds to a recording time of several seconds. This reduction of the recording time must be paid for at least partially by a worse signal-to-noise ratio.

Another method that is also based on gradient echoes was proposed by Oppelt and coauthors (1986) and termed FISP (Fast Imaging with Steady Precession). The scheme of the FISP sequence is plotted in Fig. 4.6. An essential property of FISP is that the slice gradient, the phase coding gradient and the reading gradient are always switched on a second time with equal duration but opposite sign, so that finally all spins are again in the same position they were at the beginning, that is, in phase. In this way a dynamical steady state is reached for the longitudinal as well as for the transverse magnetization. For high signal intensity the time delay between the pulses has to be shorter than the relaxation times T_1 and T_2^*. The signal is in addition to the equilibrium magnetization M_0 also proportional to

$$\frac{\sin \alpha}{(1 + T_1/T_2^*) + (1 - T_1/T_2^*) \cos \alpha} , \tag{4.8}$$

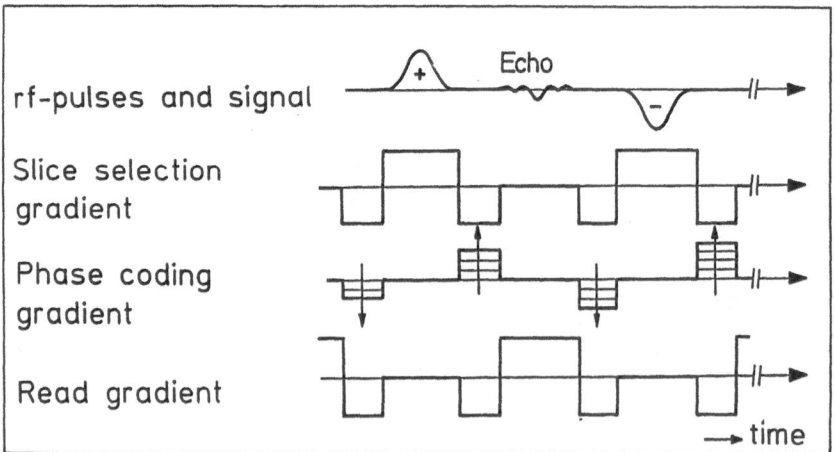

Fig. 4.6. Pulse and gradient sequence with the FISP method

which results for 90° pulses in liquids with $T_1 = T_2$ in a reduction of the magnetization M_z by a factor of 2. In general the optimum flip angle depends on the ratio T_1/T_2^*. It is

$$\alpha_{\text{opt}} = \arccos \frac{T_1 - T_2^*}{T_1 + T_2^*} \, . \tag{4.9}$$

The significant difference between FLASH and FISP is the different time-dependence of the gradients, as can be seen in Figs. 4.5 and 4.6. The unsymmetrical time-dependence of the gradients in the case of FLASH destroys all phase coherence; any remaining magnetization in the x, y plane is destroyed by a suitable gradient pulse (spoiler gradient). After several sequences an equilibrium of the magnetization M_z is reached. FLASH furnishes images that essentially reproduce the spin density ρ. By a suitable choice of the two parameters flip angle α and repetition time T_R it is possible to obtain T_1-weighted images. An extension of the FLASH-method, SNAPSHOT-FLASH makes use of the very short time required for a FLASH sequence. The spin system is prepared by placing a conventional NMR pulse sequence (for example a 180° pulse or a spin echo sequence) before the imaging sequence. This gives strongly T_1 or T_2 weighted pictures. Since all phase relations are destroyed anyway, FLASH is not quite as sensitive to unavoidable motions, for instance in the abdomen. In contrast to this, the gradient sequences of FISP are symmetric with respect to the center of the echo. In this case the transverse magnetization in the x, y plane is fully restored (aside from a decrease due to the transverse relaxation time T_2^*). After several sequences an equilibrium is reached for the longitudinal magnetization M_z as well as for the transverse magnetization in the x, y plane. Hence, one obtains with FISP an NMR tomogram that depends on T_1 as well as on T_2 and shows a very good contrast resolution in spite of the short recording time. On the other hand, FISP requires very homogeneous magnetic fields and a very precise adjustment. In their application both methods are in some respect complementary; examples are given in the following section.

Another strategy for reducing the measuring time is the RARE (Rapid Acquisition with Relaxation Enhancement) method (Hennig et al., 1984). Here the point is not to reduce the repetition time, but rather to record a series of values of the phase coding gradients simultaneously with one excitation of the spin system by different phase codings of different echoes. The difference between the normal multi-echo technique and the RARE method is explained in Fig. 4.7. In both cases, for instance eight echoes are recorded by a CPMG sequence after each excitation of the spin system. With the multi-echo technique the phase coding is identical for all eight echoes so that one obtains, depending on the ratio of the time delay between the pulses to T_2, eight different T_2-weighted echo images (Fig. 4.7a). With the RARE method all eight echoes are differently phase coded so that for the necessary total of 256 phase codings only 32 excitations instead of 256 are required; this is a time-saving of a factor of 8 (Fig. 4.7b). The fast images obtained in this manner show a medium contrast similar to the echo images 4

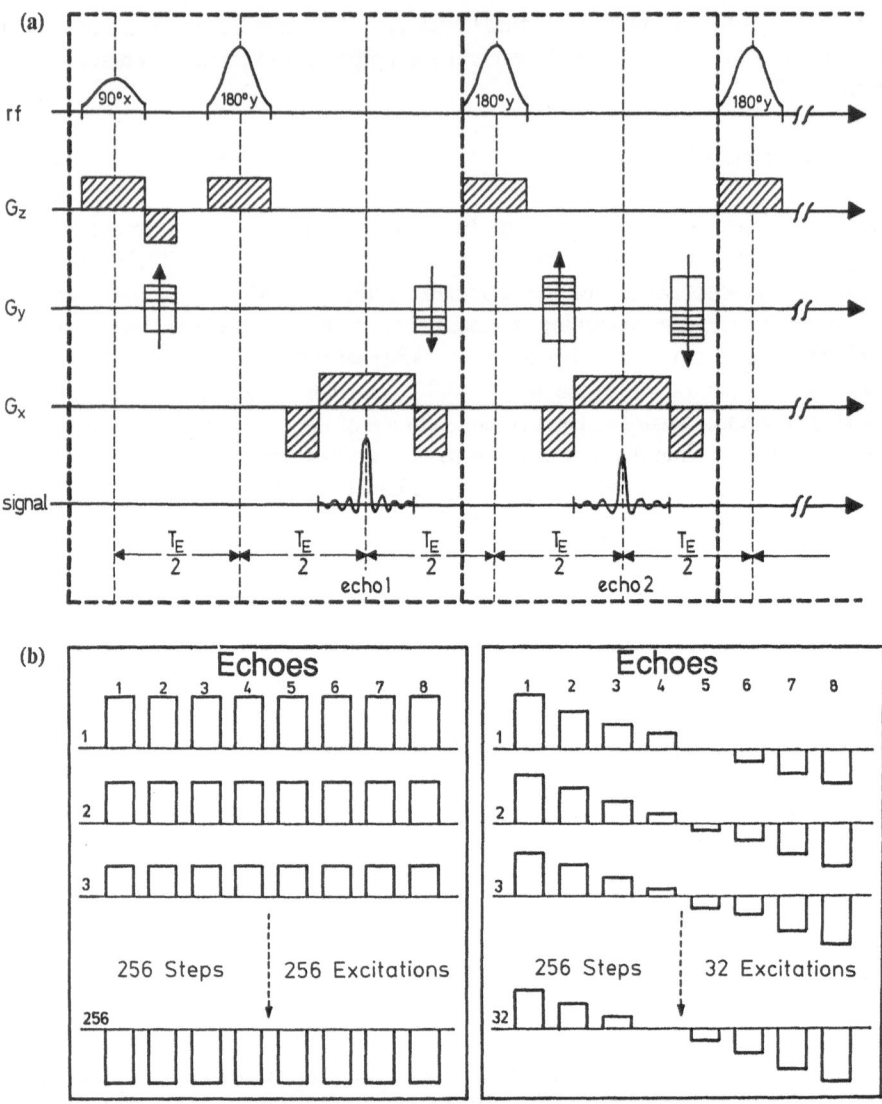

Fig. 4.7a,b. RARE imaging. (a) Pulse and gradient sequence, (b) Schematic view of the dephasing by the phase encoding gradient G_y in multiecho imaging (*left*) and in RARE imaging (*right*)

and 5 of the "normal" multi-echo series. The RARE method renders T_2 emphasized contrasts that can be varied in a well-defined way by setting the number of echoes. It is suitable for a multi-slice technique, for instance for examining the brain. However, it is unsuitable for examining moving parts of the body such as chest and abdomen because motion destroys the phase relations in the CPMG sequence. Moreover, with high field instruments the order of magnitude of the maximum tolerable rf power is easily reached because of the many 180° pulses required in a comparatively short time.

An interesting application of the RARE method depends on the strong T_2 dependence of the echo images mentioned above. For instance, the echo sequence can be chosen in a way that all signals from regions with T_2 values below 500 ms vanish, so one observes only liquid regions such as the liquor in the cerebrospinal channel. Since in this case there is no risk of superposition of various regions within the part of the body being investigated – liquid regions are rare – the slice selecting gradient can be omitted to show, so to speak, a projection in the plane of the part of the body investigated (Hennig et al., 1986). In this way one can obtain a myelogram without contrast agents when investigating the region of the neck. Since the image is a projection onto a plane, rather than a single thin slice, the entire anatomy is visible, as in an X-ray image, and the cerebrospinal channel always remains in the image plane. Such an image can, for instance, be recorded in a few seconds with 128 echoes per excitation and a time delay between echoes of 20–30 ms.

4.1.5 Three-Dimensional Imaging

Three-dimensional Fourier imaging in NMR was first discussed by Ernst and coworkers (Kumar et al., 1975). The advantages are striking:

1. The best possible signal-to-noise ratio, since with each measuring pulse the whole volume contributes to the signal, while with planar imaging only the nuclear spins of one slice produce the signal, whereas the whole volume contributes to the noise.
2. An optimum spatial resolution in all directions.
3. The possibility of examining subsequently any surface, even a non-planar one through the 3D volume investigated.

Nevertheless, 3D imaging has until recently been of little significance. This was mainly due to the very long measuring time as compared to the usual methods. Let us assume a CPMG echo sequence and comparatively favorable values: a 3D matrix of 128^3 voxels (volume elements), a repetition time $T_R = 1$ s, and a number of accumulations $A = 2$, then a measuring time of more than nine hours is required! A further technical problem was also the size of the data matrix necessary for a 3D image – for the example given above of 128^3 voxels a storage capacity of 16 megabytes is needed. However, in view of the development of improved computer technology in recent years, this is no longer a technical problem but only a financial one. The fast imaging methods discussed in the previous section make it possible to perform 3D experiments in an acceptable measuring time; in the example mentioned above the 128^3 voxels can be recorded as a 3D FLASH tomogram in about 5–10 minutes.

Fast 3D imaging is an extension of the fast 2D imaging techniques FLASH and FISP, and it uses gradient echoes as well. To obtain a 3D image, instead of a slice-selecting gradient a second phase gradient independent of the first one is coded in n steps (Johnson et al., 1983). If one, for instance, codes this second phase gradient in 128 steps like the first one, a homogeneous data set is obtained with equal resolution in all space directions. Using a computer any plane required for diagnosis can be displayed with these data. The scheme of the 3D-FLASH and 3D-FISP sequences is plotted in Figs. 4.8 and 4.9. The characteristic difference between the two is the same as was explained above for 2D imaging.

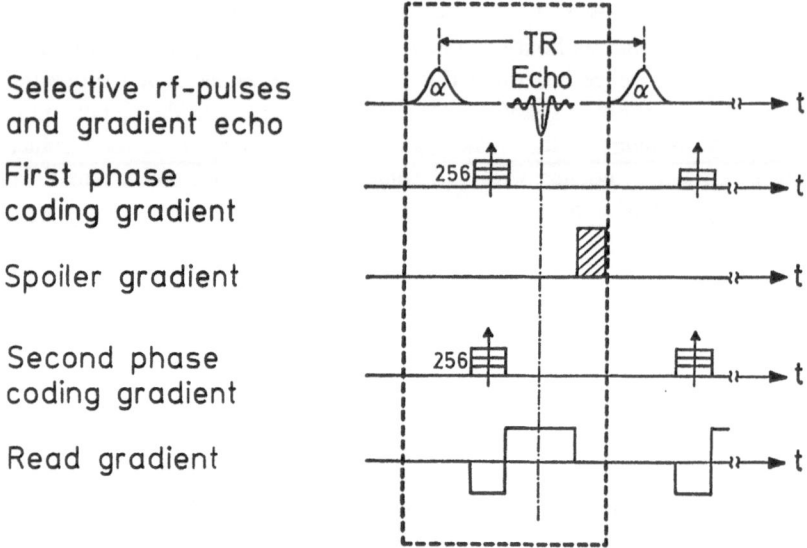

Fig. 4.8. Pulse and gradient sequence with the 3D FLASH method

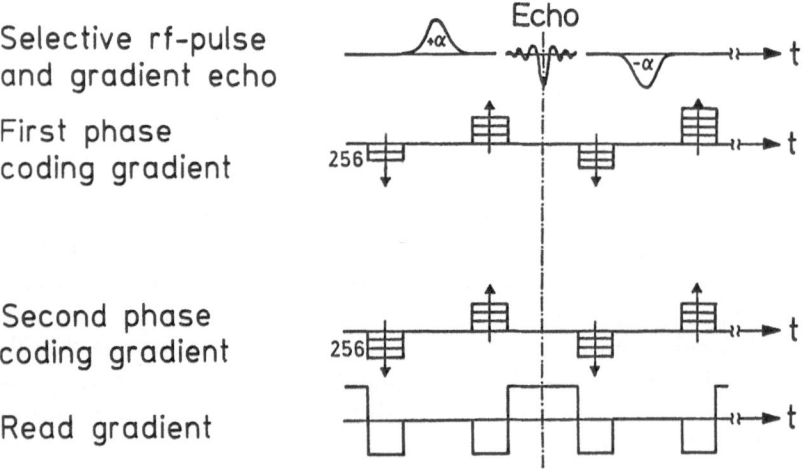

Fig. 4.9. Pulse and gradient sequence with the 3D FISP method

4.2 Some Applications of the Basic Experiments in NMR Tomography

In this section typical properties of the various imaging methods described in the preceding chapter will be illustrated with a limited number of examples.[2] Here more weight is put on the general methodological aspect than on the diagnostic details, although these are of course very relevant for practical clinical application.

4.2.1 Imaging of the Head

One of the most important applications of NMR tomography is the imaging of the head. There are a number of reasons for this. On the one hand, the skull does not hinder NMR tomography, as it does in X-ray computer tomography, since the radio frequency radiation passes practically undamped through the bones. On the other hand, recording T_1 and T_2 weighted images provides an optimum contrast resolution in the brain and the soft tissues. With patients at rest there are no motions in the head in contrast to the unavoidable motions in the thorax (heart beat, respiration) and in the abdomen (peristalsis) that make NMR imaging of these areas difficult. Hence, we have chosen the head of one of the authors (K.H. Hausser) and have recorded images in low field at 0.28 tesla and in high field at 2.0 tesla using different techniques for illustrative purposes.

Figure 4.10 shows a sagittal slice of the head recorded at 0.28 tesla (12 MHz). Note that such a slice cannot be obtained with X-ray computer tomography. The image is the central slice of a data set recorded with the CPMG multi-echo/multi-slice technique; the images a–e are differently T_2 weighted by choosing individual echoes. Figure 4.11 is a sum over all eight echoes of Fig. 4.10. In Fig. 4.12 a transverse slice through the same head at the height of the eyes is depicted. Figures 4.12a and 4.12b show again two different T_2 weighted images, Fig. 4.12c is the sum over all eight echoes and 4.12d is an image recorded with the RARE method. As we see, the RARE method produces rather good images in spite of the reduced recording time.

For comparison, the following images show the corresponding slices of the same head recorded in high field at 2.0 tesla (85 MHz). Figure 4.13 shows again four different T_2 weighted images of a sagittal slice and Fig. 4.14 the corresponding sum image. Figure 4.15 shows a transverse slice at the height of the eyes through the same head recorded with different time delays between the echoes.

The following figures are a collection of some pathological examples from the head. In all cases a matrix of 256^2 pixels is recorded from a slice of 5–10 mm

---

[2] The tomograms shown were taken at Bruker, Karlsruhe, at Siemens, Erlangen, and in the Department of Radiology of the University Hospital, Freiburg, with a Bruker BNT 1100 (Figs. 4.10–13, 4.18), a Bruker Medspec 20/100 (Figs. 4.13–15) and a Siemens Tomikon spectrometer (Figs. 4.16, 4.17, 4.19–23, 4.28).

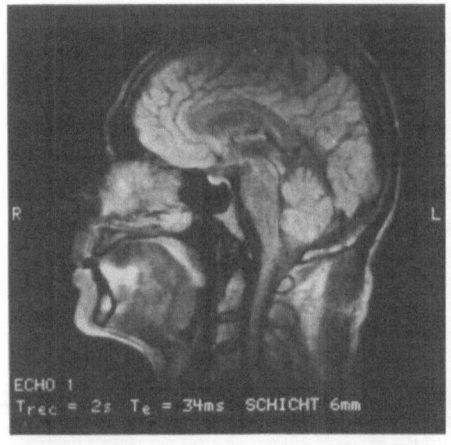

a

ECHO 1
Trec = 2s Te = 34ms SCHICHT 6mm

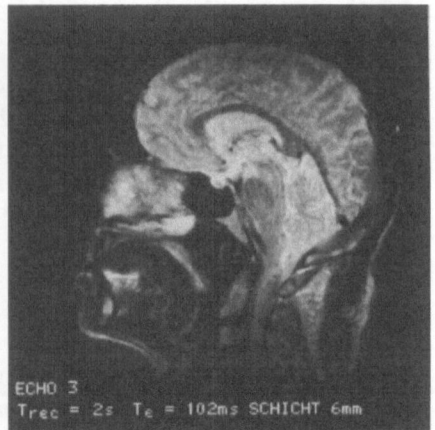

b

ECHO 3
Trec = 2s Te = 102ms SCHICHT 6mm

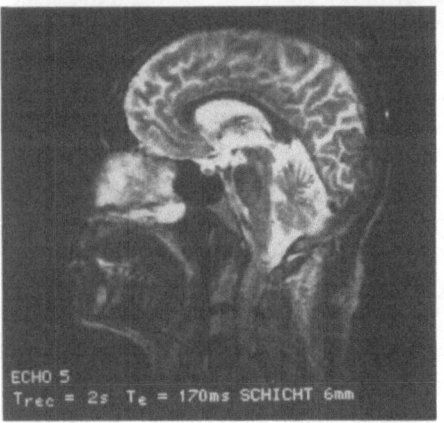

c

ECHO 5
Trec = 2s Te = 170ms SCHICHT 6mm

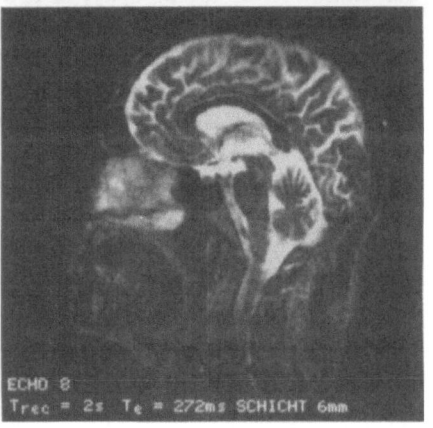

d

ECHO 8
Trec = 2s Te = 272ms SCHICHT 6mm

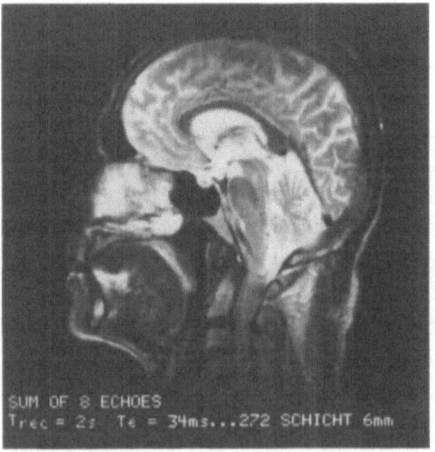

SUM OF 8 ECHOES
Trec = 2s Te = 34ms...272 SCHICHT 6mm

Fig. 4.10a–d. Sagittal cross section of a human head at 0.28 T (12 MHz). Multi-slice multi-echo technique with eight echoes of a CPMG sequence. Slice thickness 6 mm, repetition time T_R = 2s, 256^2 pixels, total recording time = 8.5 min, echo time T_E = 34 ms, differently T_2-weighted images by using different echoes: (a) first echo T_{E1} = 34 ms, (b) third echo T_{E3} =102 ms, (c) fifth echo T_{E5} = 170 ms, (d) eighth echo T_{E8} = 272 ms (Bruker, Karlsruhe)

Fig. 4.11. Sagittal cross section of a human head. Sum image of the eight echo images shown in Fig. 4.10

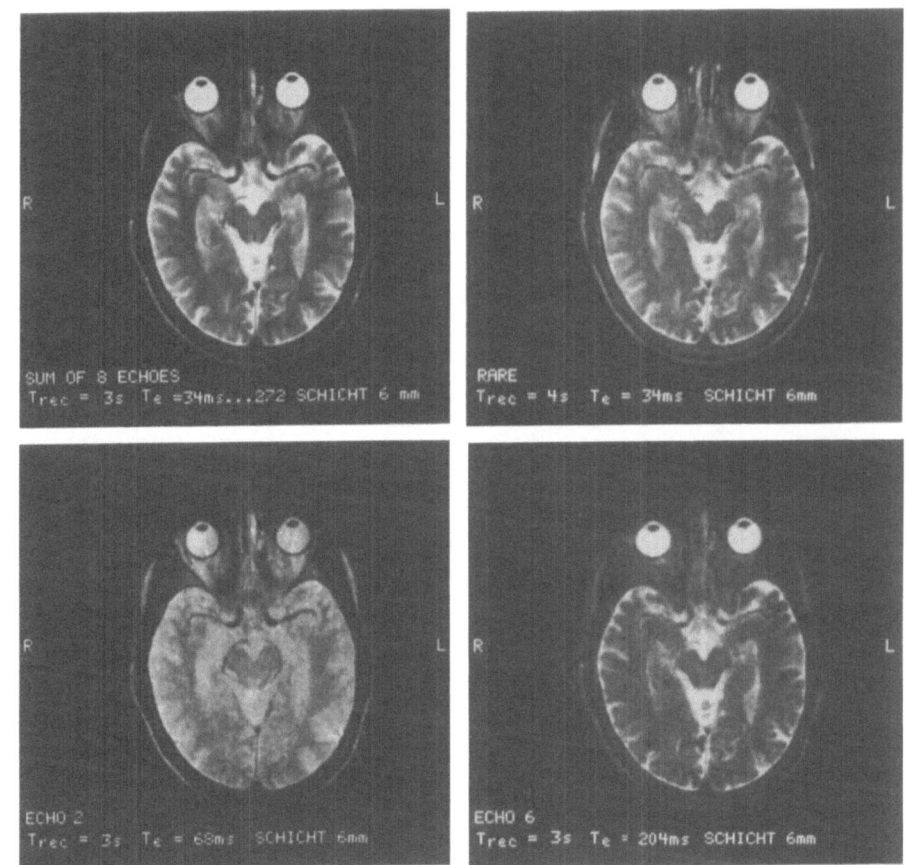

Fig. 4.12a–d. Transverse cross section of a human head. T_R = 3 s, total recording time approx. 13 min. (a) Second echo T_{E2} = 68 ms, (b) sixth echo T_{E6} = 204 ms, (c) sum image of eight echoes, (d) same as a-c, RARE method, T_R =4 s, eight echoes differently phase coded, total recording time about 100 s. Other data as in Fig. 4.10 (Bruker, Karlsruhe)

thickness (the exact thickness of the slice is given with each picture). In the case of the head this corresponds to a linear resolution in the plane of about 1 mm.

Figure 4.16 shows a craniopharyngioma recorded with the spin echo technique as a sagittal slice at 1.0 tesla. By comparing two echo images with different delay times T_E between the echoes, the possibilities of different contrast resolution by different T_2 weighted images can clearly be recognized. Comparing the two images enables one to distinguish between tumor and accompanying edema.

Figure 4.17 shows a cystic tumor in the brain, also recorded at 1.0 tesla with the spin echo technique; as in the preceding image a very marked change in the contrast resolution occurs that is due to a different T_2 weighting produced by a factor of 4 difference in the delay time T_E of the echoes. Since the cystic tumor consists essentially of liquids with long T_2, it shows a strong contrast with respect to its surroundings after a longer time delay between the echoes.

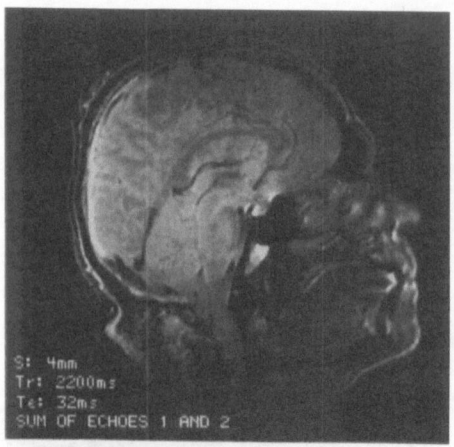

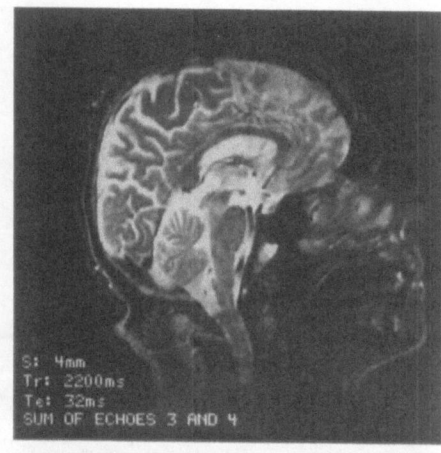

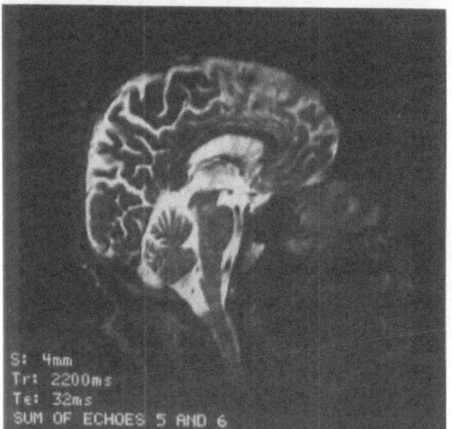

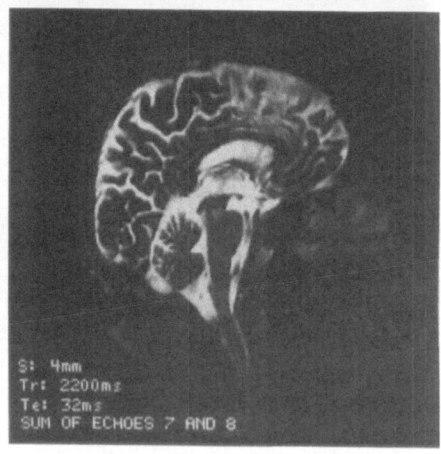

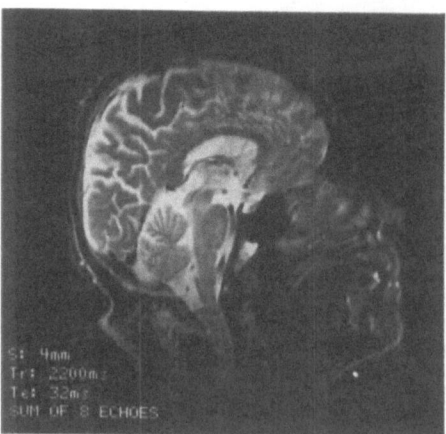

Fig. 4.13a–d. Sagittal cross section of a human head at 2.0 T (85 MHz). Multi-slice multi-echo technique of a CPMG sequence, slice thickness 4 mm, repetition time T_R = 2.2 s, 256^2 pixels, total recording time about 9 min. Echo time T_E = 32 ms, differently T_2 weighted images by using different echoes: (a) First plus second echo, (b) third plus fourth echo, (c) fifth plus sixth echo, (d) seventh plus eighth echo. The images of Figs. 4.13 to 4.15 were obtained with a superconducting magnet (Bruker, Karlsruhe)

Fig. 4.14. Sagittal cross section of a human head. Sum image of the eight echo images shown in Fig. 4.13 (Bruker, Karlsruhe)

152

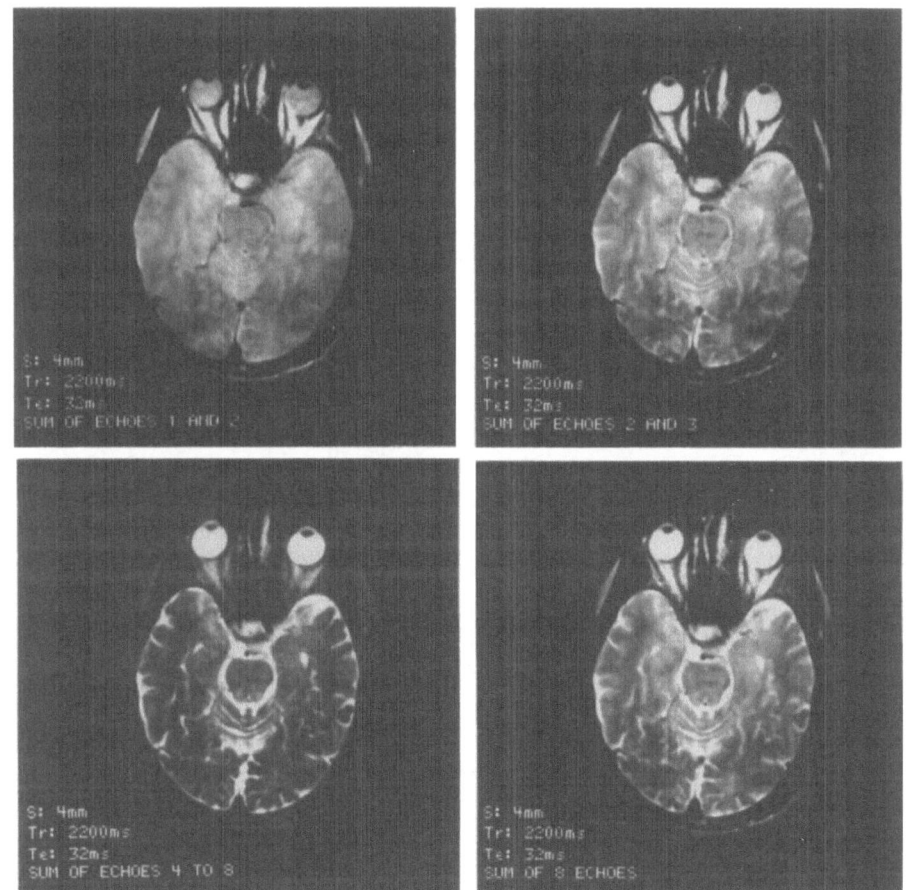

a

b

c

d

Fig. 4.15a–d. Transverse cross section of a human head. (a) First plus second echo, (b) second plus third echo, (c) sum of the fourth to eighth echoes, (d) sum of all eight echoes. Other data as in Fig. 4.13 (Bruker, Karlsruhe)

Figure 4.18 shows again a transverse slice through the head of a patient who suffers from multiple sclerosis. The brighter parts, which are regions of higher signal intensity, correspond to multiple sclerosis lesions; they are particularly marked in the second echo image of the differently T_2 weighted images of a CPMG echo sequence.

4.2.2 Imaging in Other Parts of the Body

A multitude of applications of NMR tomography is also possible in other parts of the body where whole body images with the same data matrix of 256^2 pixels lead to a somewhat reduced spatial resolution between 1 mm and 2 mm because of the larger diameter of the body. Problems arise in the thorax (heart beat)

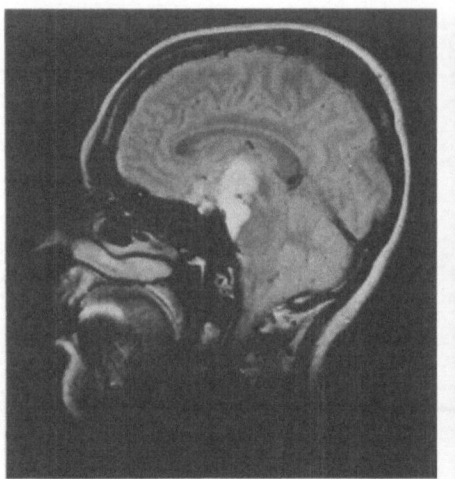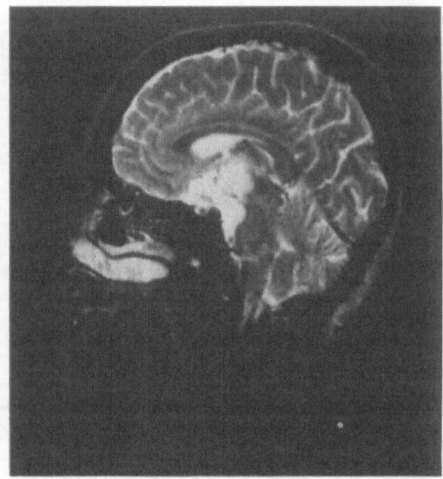

a   b

Fig. 4.16a,b. Craniopharyngioma. Sagittal cross section recorded at 1.0 T (42.6 MHz) with different echo times. (a) CPMG spin echo technique, slice thickness 6 mm, $T_E = 28$ ms, $T_R = 3.30$ s, total recording time for the multi-slice image about 15 min. (b) Same as (a), but $T_E = 112$ ms (Siemens, Erlangen)

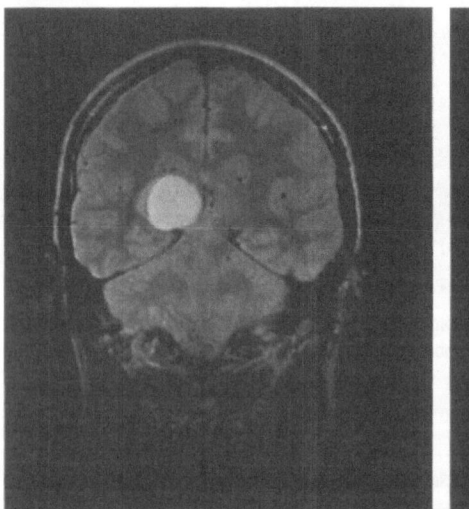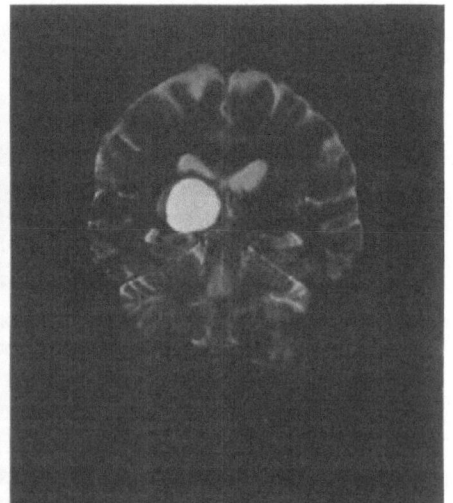

a   b

Fig. 4.17a,b. Cystic tumor. Frontal cross section at 1.0 T, CPMG spin echo technique, slice thickness 5 mm, $T_R = 3.20$ s. (a) $T_E = 30$ ms, (b) $T_E = 120$ ms (Siemens, Erlangen)

and in the abdomen (peristalsis) due to motions. In the case of periodic motions triggered images can be obtained with good resolution. Figure 4.19 shows an EKG triggered image of the heart at 1.0 tesla where the anatomical details can be well recognized. Note the coronary arteries marked by an arrow. Figure 4.20 is again an EKG triggered image of the thorax at 1.0 tesla that shows clearly an aortic aneurism.

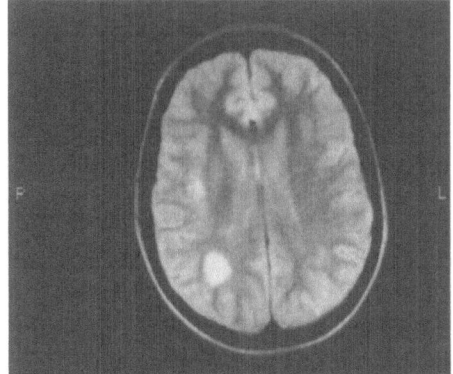

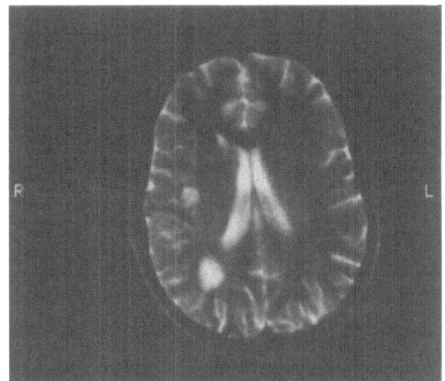

a

b

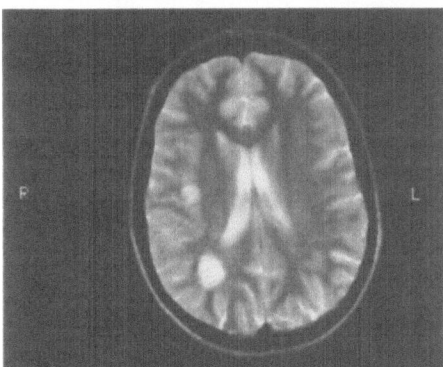

c

Fig. 4.18a–c. Multiple sclerosis. Transverse cross section of the brain at 0.28 T. CPMG spin echo technique, slice thickness 6 mm, echo time T_E = 34 ms, (a) First plus second echo, (b) fifth plus sixth echo, (c) sum image of all eight echoes (Department of Radiology, University of Freiburg)

Fig. 4.19. Transverse cross section of the thorax at the height of the heart at 1.0 T. Spin echo technique. EKG-triggered, slice thickness 10 mm, T_R = 0.82 s, T_E = 17 ms. (*Arrow*) coronary arteries (Siemens, Erlangen)

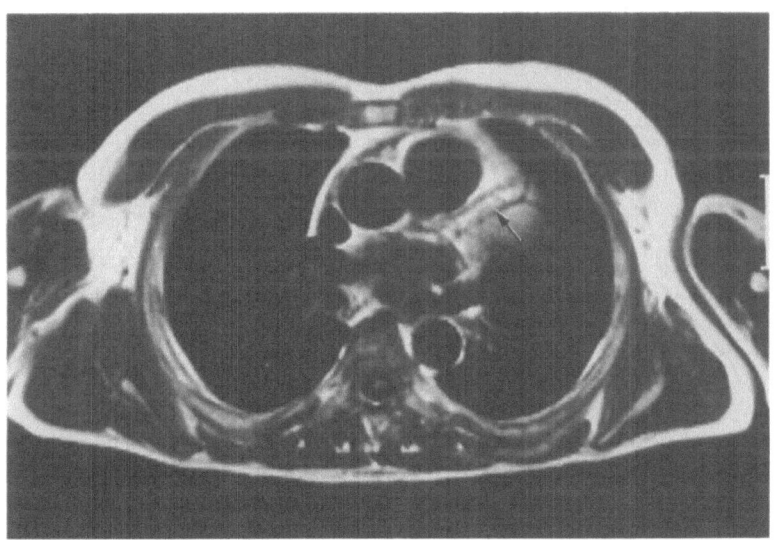

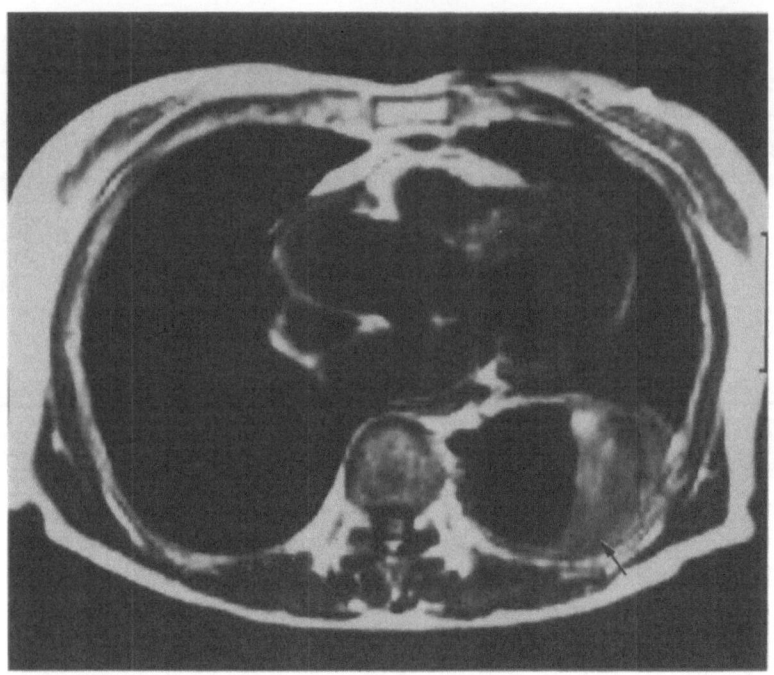

Fig. 4.20. Transverse cross section of the thorax at 1.0 T. Spin echo technique, EKG-triggered, slice thickness 8 mm, $T_R = 0.7$ s, $T_E = 28$ ms. (*Arrow*) aneurism of the aorta (Siemens, Erlangen)

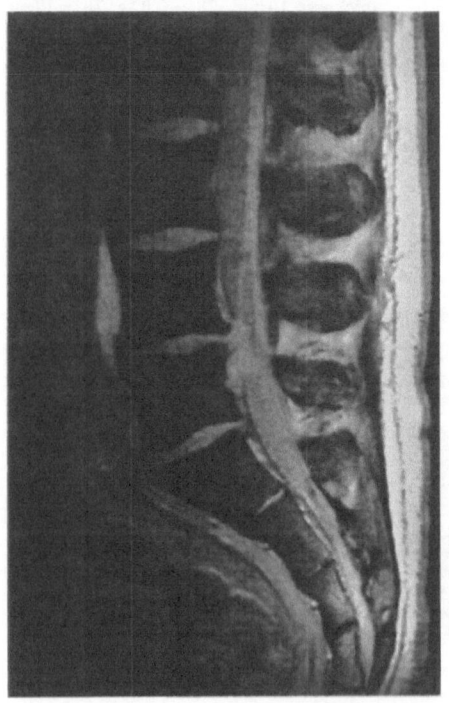

◄ **Fig. 4.21.** Sagittal cross section in the area of the lumbosacral spine at 1.0 T. Fast imaging method FISP, slice thickness 5 mm, $T_R = 0.1$ s, $T_E = 13$ ms. The slipped disk in the center of the image can be clearly recognized (Siemens, Erlangen)

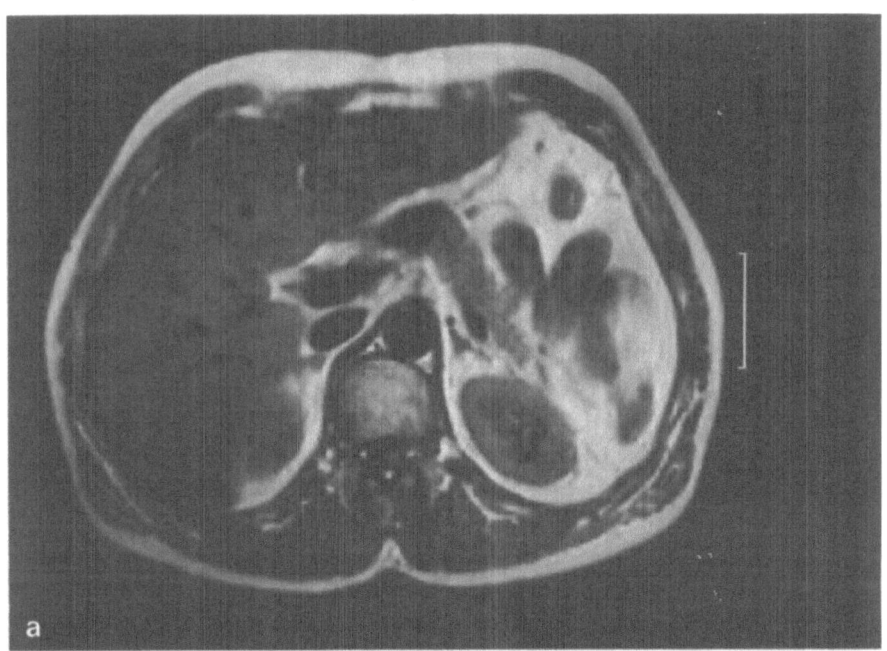

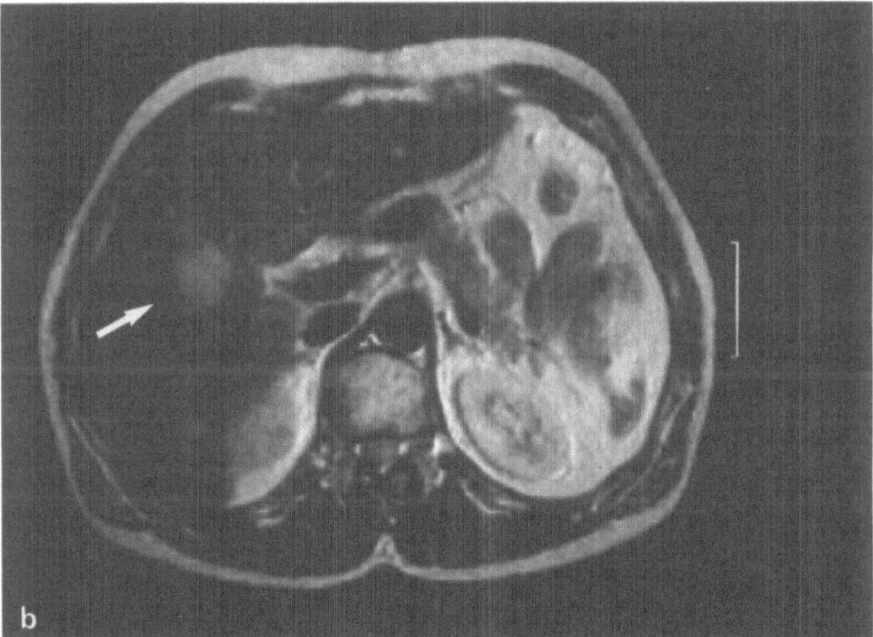

Fig. 4.22a,b. Transverse cross section of the abdomen at 1.0 T. Spin echo technique, slice thickness 8 mm, $T_R = 0.8$ s, (a) $T_E = 28$ ms, (b) $T_E = 70$ ms. Metastases of the liver can be recognized in (b) (*arrow*) but not in (a) (Siemens, Erlangen)

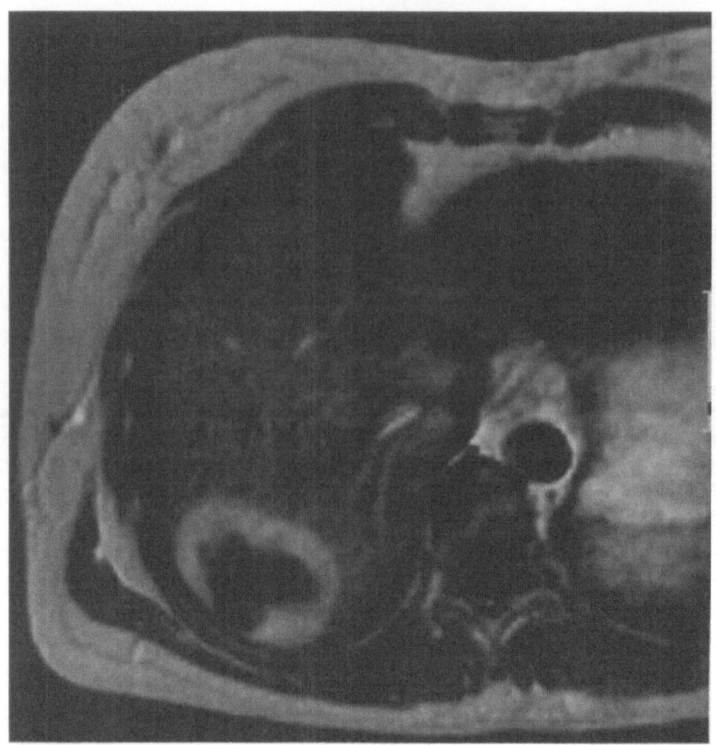

Fig. 4.23. Transverse cross section of the abdomen at 1.0 T. Spin echo technique. Slice thickness 10 mm, $T_R = 0.8$ s, $T_E = 30$ ms. An echinococcus cyst can be recognized in the left lower part of the image

NMR tomography is also quite suitable for images of the spine. Figure 4.21 shows a slipped disk in the lumbosacral spine. An exact localization, as the basis for an eventual surgery, is most conveniently obtained non-invasively with NMR, which can display sagittal slices easily, in distinction to X-ray computer tomography. At the same time Fig. 4.21 is an example of a recording mode with the fast image method FISP; the total recording time was about 5 s.

Figure 4.22 is a transverse slice of the abdomen; both images a and b are T_2 weighted by different delay times T_E between the echoes. In this case it is a patient with liver metastases, which can be seen clearly in Fig. 4.22b with a long delay time of the echo, $T_E = 70$ ms, while they cannot be recognized in Fig. 4.22a with a shorter delay time $T_E = 28$ ms.

Figure 4.23 is the image of a patient with a taeniasis; the echinococcus cyst in the liver can be clearly recognized in the left lower part of the image.

4.3 Special Applications of NMR Tomography

In addition to the spin density and the relaxation times, there are still other effects that may influence the NMR signal, for instance motional processes and the splitting of the NMR signals into several lines owing to the chemical shift. In general, one tries to suppress these influences on the NMR image because they produce unwanted artifacts. On the other hand, we shall see that they can be used as sources of new information. In recent years a major part of the technological development was devoted to the construction of NMR systems by which images of large objects could be recorded. An interesting new development of NMR tomography is NMR microscopy (micro imaging) that concentrates on investigating very small objects in high resolution, but uses in principle the same methods already described.

4.3.1 Chemical Shift in Imaging

As already mentioned, the NMR signal from tissues is essentially determined by the water content because of its concentration. In principle, all other protons of sufficiently mobile molecules contribute to the signal as well; however, their concentrations and hence the corresponding signal intensities are in general lower than those of water by several orders of magnitude. An important exception are the lipids in fatty cells that may occur in very high concentrations. The resonance lines of the methyl protons of the lipids are shifted by about 3.5 ppm to high field with respect to the water signal. The absolute chemical shift measured in frequency units is proportional to the external magnetic field B_0 (1.3). At a low field of about 0.2 tesla the water and the fatty signal are about 30 Hz apart, which means that they can hardly be separated. The situation is different at the high field of 2 tesla where the resonance lines are separated by about 300 Hz and may cause considerable perturbations of the image. For instance, a slice selection with a frequency selective pulse renders a selection of two slices shifted with respect to each other. Methods can be sought to avoid these perturbations; however, the possibility of separating the images of fat and of water protons and the production of pure fat and pure water images, can be of interest from the diagnostic point of view, for instance for distinguishing lymph nodes from surrounding fatty tissue in mammography.

In principle, two completely different ways for separating water and fatty protons based on the different chemical shifts were chosen. With one of these methods, called by its authors CHESS (CHEmical Shift Selective) tomography (Haase et al., 1985), a certain imaging pulse sequence acts only on one of the two types of protons. This can again be achieved in two ways. One way is to saturate one of these proton types by applying a frequency selective radio frequency pulse without a gradient and subsequently perform the pulse sequence as usual, where, of course, only the other type of proton is measured because the magnetization of the 'saturated' type is zero (Haase et al., 1985). This method requires a homogeneity of the B_0-field that is considerably better than 3.5 ppm, a condition which is in general satisfied since a good magnet has a homogeneity of 1 ppm or below over the volume of the body being investigated.

In another variant of this method, one type of proton is not excluded by saturation, but the desired resonance line is positively chosen by selective excitation, which is irradiated together with a radio frequency pulse for selecting the plane of the image. This variant of CHESS tomography can be realized, for instance, with a series of three pulses that produce a stimulated echo (Haase and Frahm, 1985). As early as 1950, such stimulated echoes were described by Hahn (Hahn, 1950); they occur after applying a sequence of three radio frequency pulses of the type $90°$-τ_1-$90°$-τ_2-$90°$-τ_3 at time $\tau_3 = \tau_1$, in addition to the usual spin echoes.

The radio frequency pulse and the gradient sequence is plotted in Fig. 4.24. For the first two echoes frequency-selective pulses are used that can be realized in several different ways; for instance, by a Gaussian distribution of the amplitude of the radio frequency with a pulse length of about 10 ms that corresponds to a Gaussian frequency distribution with a half-width of about 100 Hz. In this respect the pulses are to a large extent exchangeable; for instance, it may be advantageous to make the third pulse instead of the second frequency-selective for slice selection. The advantages of this technique are:

1. No problems with eddy currents because no gradients are switched on before the first pulse.
2. The experiment does not depend critically on the precise field angle or on the homogeneity of the B_1-field and can, therefore, even be performed with surface coils.
3. By varying the techniques described, images of several NMR resonance lines, either of one or of several substances, can simultaneously be obtained.

A variant of the CHESS technique was described (Hennig et al., 1986), where one of the two magnetizations is inverted by a CHESS $180°$ pulse and subsequently a normal 2D FT experiment is performed, for instance, with a CPMG spin echo sequence. In this case it is appropriate to invert the magnetization with a longer T_1 to minimize falsifications of the amplitude by T_1 relaxation. Subsequently the experiment is repeated without the inversion pulse. The sum of both results

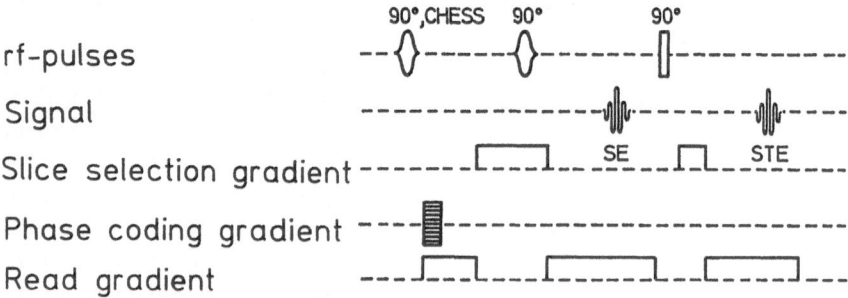

Fig. 4.24. Pulse and gradient sequence with CHESS (chemical shift selective) tomography

renders again a pure tomogram of the fat protons, and the difference provides that of the water protons.

A totally different method proposed by Dixon (1984) called chemical shift imaging (CSI) is based on the fact that the fat and water proton spins rotate in the x, y plane after a 90° pulse with different angular frequencies ω^F and ω^W. This causes a phase difference that increases linearly with time. However, this phase difference decreases again to zero in the case of a normal 90° – τ – 180° echo pulse sequence after the 180° pulse at the time of the echo, so that the magnetization of the fat protons M^F and of the water protons M^W at the time of the echo are again in phase, and the sum of their absolute values $M^F + M^W$ contributes to the echo. However, if one applies an additional gradient to the spin system during the refocusing time, the refocusing of the two different magnetizations M^F and M^W no longer occurs at the same time. Hence, one obtains, depending on whether the spin echo and the gradient echo coincide, either the same sum image of $M^F + M^W$ as without the bipolar gradient, or, if they differ by half a period $\pi/(\omega^F - \omega^W)$, a difference image $M^F - M^W$. Adding these two images results in a pure tomogram of the fat protons and subtracting them provides the same for the water protons.

4.3.2 Flow Effects in NMR Tomography

The human organism is not an immobile, rigid system; even at rest it has a number of internal motions. Continuous flow motions are of special importance for an adequate function of the organism. The most important of these processes is the blood flow in the blood vessels, but in addition there are flow phenomena of other types, such as the flow of lymph in the lymphatic system, the flow of urine in the ureters and the flow of the liquor in the cerebrospinal system. For the oriented motion of liquids there are two important types of flow: laminar flow and turbulent flow (Fig. 4.25a). Turbulent flow is typical for relatively high flow velocities as they may occur in the arterial system. Laminar flow is typical for low speeds of flow in vessels of relatively large diameters such as those found in the veins. It shows often a parabolic speed profile (Fig. 4.25b). With plug flow (that may be either laminar or turbulent) the speed of flow is identical over the whole cross section of the vessel (Fig. 4.25c). It is very easy to describe and is therefore frequently used as a simple model, but it does not play an important role in human organism. If one compares the signal intensity of a flowing liquid with an immobile one, the flow effects may lead to either a decrease or to an increase of the signal intensity in the NMR tomogram. The basic phenomena can best be described by a plug flow perpendicular to the image plane.

In an experiment that is repeated many times the repetition time is usually chosen not shorter than three times the spin-lattice relaxation time in order to optimize the signal-to-noise ratio. For shorter repetition times slice-selective 90° pulses lead to a partial saturation of the signal of the liquid. If additional unsaturated liquid flows continuously into the slice, the saturation is partially cancelled

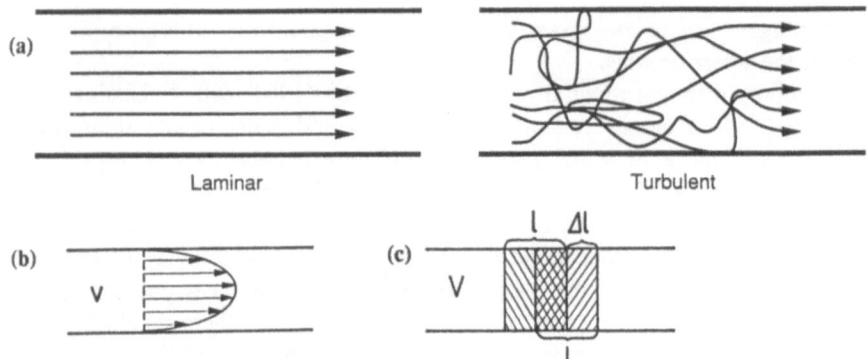

Fig. 4.25. (a) Streamlines in laminar and turbulent flow. (b) Typical parabolic velocity profile in laminar flow. (c) Plug flow, a volume V element of length l moves during a time t with a velocity v for a distance Δl

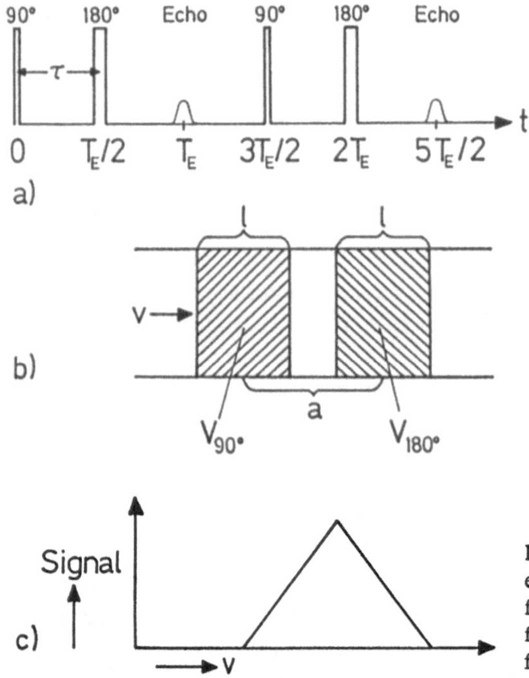

Fig. 4.26a–c. Flow effects with the spin echo method. (a) Pulse sequence for the flow method. (b) Conditions with plug flow in a tube. (c) Signal intensity as a function of the velocity v with plug flow

and a stronger signal is obtained. The situation is somewhat more complicated if a sequence of several radio frequency pulses is applied and the slice is defined by the first excitation pulse. Let us assume that a cylinder of a length l is excited by the first frequency-selective 90° pulse of a 90° - τ – 180° echo pulse sequence (Fig. 4.26). The excited liquid volume V is in this case $V = r^2 \pi l$. If in the time $\tau = T_E/2$, the distance between the 90° and 180° pulse, the liquid flows with the velocity v for a distance $\Delta l = v T_E/2$, the volume V', on which both

162

pulses act, is reduced by $r^2\pi\Delta l$, that is, $V' = r^2\pi(l - \Delta l)$, and accordingly the signal is reduced in the ratio V'/V (Fig. 4.25). It follows that with this assumption the signal can be reduced to zero with a sufficiently high speed of flow if $\Delta l = v T_E/2 \geq l$.

The dependence on the signal amplitude just described is only valid if there is no saturation, that is, if the repetition time T_R is much longer than the relaxation time. For shorter repetition times the effect of an increase of the signal through the flowing of unsaturated spins into the excited volume described above has to be taken into account as well. For instance, if the next pulse sequence begins at $T_E/2$ after the echo ($T_R = 3T_E/2$), a volume $V'' = r^2\pi\cdot 3\Delta l$ contains a new inflow of unsaturated spins and accordingly contributes to the signal intensity as strongly as with the first pulse sequence reduced by a volume factor V''/V', whereas the remaining volume $V - V' = r^2\pi(l - 3\Delta l)$ contributes considerably less to the total signal intensity because of the partial saturation. It is not difficult to see that with a certain speed of flow v, which depends on the degree of saturation, that is, on the ratio T_1/T_R, these two effects may just compensate each other and by chance the same signal may result as with the immobile liquid.

If one wants to obtain a maximum signal in spite of the speed of flow v, this can be achieved by shifting the volume V_{180}, on which the 180° pulse acts, by the distance a with respect to the volume V_{90}, which is excited by the 90° pulse by utilizing a frequency shift in the presence of the gradient in the flow direction (Fig. 4.26). In this case no signal is observed for $a > l$ with an immobile liquid. With a flowing liquid the signal intensity is a function of v; it is zero for $vt = a - l$, then increases linearly until $vt = a$ and decreases again until $vt = a + l$ to zero, as is plotted in Fig. 4.26.

We have already mentioned that the simplifying assumption of a plug flow for all spins is unrealistic. The flow of blood in the vessels can be either turbulent or laminar. Let us consider, with reference to Fig. 4.26, the more realistic case of laminar flow (Fig. 4.27). In this case the speed of flow increases continuously from the walls to the center, and the profile of the speed of flow is often parabolic. In the example chosen for Fig. 4.27 only the spins in the cross-hatched volume are excited both by the 90° pulse and by the 180° pulse and consequently contribute to the signal.

As we have seen, no signal is obtained if, for instance, the 90° and the 180° pulses of a 90° - τ - 180° echo pulse sequence act on different volumes V and V'. If, however, part of the spins in volume V is transported during the time between the 90° and the 180° pulse from volume V to volume V', for example

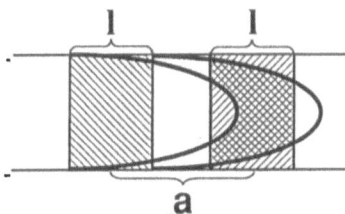

Fig. 4.27. Flow effects with laminar flow

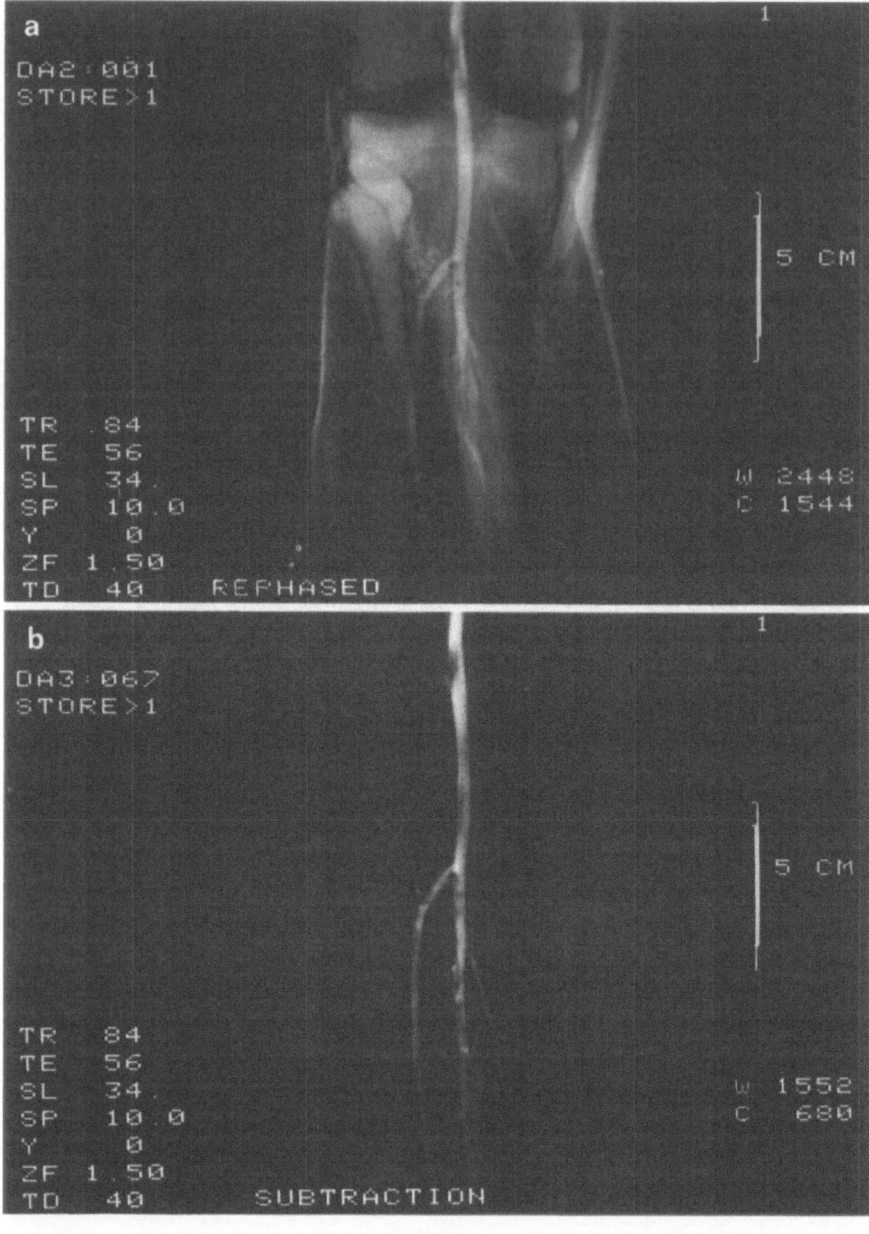

Fig. 4.28a,b. Image of the blood vessels in the region of the knee at 1.0 T. Modified spin echo technique (Laub and Kaiser, 1988) where flow effects are used for increased or decreased contrast in the image of the blood vessels. Slice thickness 10 mm, repetition time $T_R = 0.84$ s, echo time $T_E = 56$ ms. (a) Flow increased image. (b) Difference image between flow increased and flow decreased image, so that only the vessels with flowing blood are visible (Siemens, Erlangen)

by the flowing blood, an NMR signal is obtained. This principle can be used in several ways for selectively imaging blood vessels in the NMR tomogram (Fig. 4.28).

The speed v with which a homogeneous liquid flows through a rigid tube can be measured with NMR by utilizing the reduction in the spin-spin relaxation time T_2. If a liquid flows through a tube with the speed v, measurement of the spin-spin relaxation time T_2 in a slice or a short cylinder with a CPMG pulse sequence results in a shorter value T_2 as compared to a liquid without flow. Quantitatively, the relation exists

$$\frac{1}{T_2^*} = \frac{1}{T_2} + \frac{v}{l} \quad , \tag{4.10}$$

where l is again the thickness of the slice, i.e. the length of the cylinder. Since T_2, T_2^* and l can be measured directly, the speed v can be determined using (4.10). Alternatively, one can also measure T_2^* for different values of l and then determine v using (4.10) without knowing T_2. If the speed of flow v and the cross section of the tube Q are known, the quantity Qv can be calculated that flows through the tube per time unit; thus, for instance the cardiac output can be measured by NMR.

So far we have tacitly assumed an even flow with constant speed v. This assumption is not valid for the flow of blood in veins and is completely unrealistic in arteries; in the latter the flow strongly pulsates with the heart beat. An EKG-triggered CPMG pulse sequence, for instance in the aorta, does not render a signal intensity of the individual echoes continuously decreasing with the time constant T_2, but gives very strongly varying T_2^* values in the course of the pulse cycle, from which the speed of flow v in the different phases of the pulse cycle can be derived.

To determine in this case the amount of blood that flows through the vessel per time unit, the product Qv has to be replaced by the integral

$$\frac{Q}{t_2 - t_1} \int_{t_1}^{t_2} v(t) dt \tag{4.11}$$

where the time $t_2 - t_1$ is equal to the length of a pulse cycle. However, a quantitative evaluation of such a measurement for determining the amount of blood that flows through a certain vessel is very problematic because it is based on several simplifying assumptions, which are not all justified for the blood circulation in the human body. First of all, blood is not a homogeneous liquid, but a two-phase system of blood plasma and erythrocytes. The T_1 and T_2 values depend on a number of parameters, such as viscosity, temperature, pH value and content of oxygen. Furthermore, blood vessels are not rigid tubes, but have elastic walls, where differences of pressure are propagated in a wave-like manner. For these reasons measurements of the amount and speed of blood yield only approximate values with large ranges of errors at the present state of technical development.

So far we have restricted ourselves to investigations of flow perpendicular to the image plane because in this case the principle of flow of a homogeneous liquid in a rigid tube can be described comparatively easily. The situation is considerably more complex when describing the flow in the image plane and even more so in the general case when the direction of the flow is at an arbitrary angle with the image plane. This case has also been investigated theoretically and experimentally, but a detailed discussion would go far beyond the scope of this introduction. We therefore refer to the literature; a rather detailed review with more than 100 references was given by van As and Schaafsma (1986). A general problem when investigating blood flow in arteries and veins is the strong signal from the surrounding, immobile tissue. However, this perturbing signal can be largely suppressed with suitable methods (Hennig et al., 1988).

In spite of the complexity of the theory, investigations of flow are already of great diagnostic importance. Even in parts of the body not accessible to ultrasound investigations, disturbances of the blood circulation by a total or partial blocking of large vessels can be non-invasively detected by NMR.

4.3.3 Contrast Agents in NMR Tomography

One particular strength of NMR tomography is that for different tissues a very good contrast resolution can generally be obtained by variations of the parameters depending on the different relaxation times T_1 and T_2, and that this can be achieved non-invasively without applying contrast agents. What reason is there to use contrast agents nevertheless? It turns out that this is useful in cases where the differences between the T_1 and T_2 values of neighboring tissues is so small that they cannot be distinguished from the variations of these parameters. Neighboring tissues, such as a tumor and an edema beside it, sometimes cannot be distinguished because they have the same long relaxation times, but can often be observed separately by using contrast agents.

Contrast agents are all paramagnetic and therefore they shorten T_1 as well as T_2 (see Sect. 1.3.7). The contrast agents aplied so far are mostly chelates of rare earths, in particular, the gadolinium-diethylene-triamin-pentaacetic acid complex (Gd-DTPA). However, there are already some results with other contrast agents, like small ferrites and stable free nitroxide radicals.

A restriction in applying Gd-DTPA is the fact that it remains in the intravascular and intercellular compartments and cannot penetrate cell walls; nor can it pass the blood-brain barrier. In particular, this permits an important diagnostic application: If the blood-brain barrier is damaged, for instance by a trauma, by an infarct, by a tumor or by drugs, it becomes permeable to contrast agents that can then be enriched, for example, in a tumor. Injuries to the blood-brain barrier are frequently connected with edema that can make the detection of small metastases in a normal NMR tomogram difficult or even impossible; here Gd-DTPA can provide the necessary contrast. Gd-DTPA is also very suitable for determining the size and the precise dimensions of a tumor. One assumes that the paramagnetic contrast agent is enriched in the tumor, but not in the neighboring tissue, which increases the contrast between the tumor and its surroundings.

Other organs, such as kidney, liver and spleen, have been investigated by using Gd-DTPA contrast agents. In connection with fast imaging methods like FLASH, the absorption of Gd-DTPA, for example in a liver tumor, can be observed as a function of time after its application.

Stable nitroxide radicals have for a long time been used in biological investigations. However, their application in NMR tomography has so far not been widely investigated. Nevertheless, they are of potential interest for two reasons: First of all, they can be easily bound to carrier molecules that are enriched preferentially in particular tissues only. Secondly, they can penetrate the cell walls, so that one can hope to obtain more information on metabolic processes within the cell. However, the application of nitroxide radicals in NMR tomography is still at its beginning.

A crucial question is, of course, whether contrast agents are innocuous or toxic. While the element Gd itself is toxic, so far no negative side effects of Gd-DTPA have been found. Only an insignificant reversible change of the serum-iron concentration and the iron binding capacity was reported. The secretion of Gd-DTPA by the kidneys with a time constant of 20 minutes is also an indication of its non-toxicity.

4.3.4 NMR Microscopy

In this section we shall discuss the maximum obtainable spatial resolution in 2D and 3D imaging of small objects. Here we use the expression NMR microscopy or micro-imaging exclusively for the investigations of small objects if the spatial resolution achieved is higher than that one of the human eye, namely about 50μm. If one uses 128^2 pixels in a plane or 128^3 pixels in three-dimensional space, this includes objects of a dimension up to 5 mm. For investigating larger objects with a lower resolution the expression mini-imaging is occasionally used in distinction to macro-imaging. However, we shall not discuss this in greater detail because it is in principle not different from the two extremes, macro and micro imaging. The obtainable resolution of an NMR microscope is limited by the maximum possible gradient strength and above all by the signal-to-noise ratio (Laukien, 1984; Kuhn, 1989).

The gradient strength that is required depends on the linewidth and on the frequency range within which the resonance frequencies of the investigated nuclear spins are found as a result of their chemical shifts; we restrict ourselves to ^{1}H nuclei because of their favorable signal-to-noise ratio. The linewidth is, neglecting factors which depend on the lineshape, of the order of $1/T_2$. In liquids this requires, for typical T_2 values of $10^{-2} - 1$ s and a resolution of 10μm, gradient strengths of several mT/cm, which can be produced without difficulty at the present state of technical development. In solids with typical T_2 values of 10^{-5} s, however, NMR microscopy is only possible if large technical investments are made, such as MAS magic angle spinning and synchronized rotating gradient fields.

Because of different resonance frequencies owing to chemical shifts, it is in general neither possible nor reasonable to use all ^{1}H nuclei for imaging. It is, therefore, advisable to restrict oneself to one type of nuclear spin with uniform chemical surroundings, usually to the ubiquitous ^{1}H nuclei of water, the chemical shift of which does not extend over more than 0.5 ppm. In the range of one pixel the gradient has to be larger than this field range; this corresponds, with 11.75 tesla (500 MHz) and a resolution of 10μm, to a gradient strength of about 6 mT/cm, which nowadays can be attained without difficulty for the small dimensions required for NMR microscopy.

The situation is different with the signal-to-noise ratio. This is not surprising considering the fact that NMR detection is a comparatively insensitive procedure because of the small magnetic moments of the atomic nuclei, and because of the small population difference between the energy levels involved at physiological temperatures. With the present detection technique, a minimum amount of about 10^{15} protons in a sample is required in a magnetic field of 11.7 tesla (500 MHz) for a signal-to-noise ratio of 10 : 1. This corresponds in most biological tissues to a cube of 20μm edge length, and accordingly this is the maximum obtainable resolution at present. If one restricts oneself in the third dimension to a somewhat reduced resolution of about a factor of 2, this corresponds to a voxel of the dimension $20 \times 20 \times 40\mu$m^3. The extension in the third dimension is frequently made much larger, several hundred micrometers up to one millimeter; in our opinion it is then inappropriate to define the resolution on the basis of only the other two dimensions.

What was stated in Sect. 4.1 for the more favorable signal-to-noise ratio for images in 3D techniques holds of course for NMR microscopy as well. To reach a resolution higher than 20 μm, any improvement requires great effort. To gain a factor of 2 in linear resolution in 3D space, an increase of the signal-to-noise ratio by a factor of $2^3 = 8$ is necessary, which is almost one order of magnitude. Since the noise of the sample is small because of its small size, one could try to reduce the noise at the input of the amplifier by cooling the receiver coil down to helium temperature. This is, however, very problematic for biological investigations because of the large temperature difference between the sample and the receiver coil.

Even if one succeeds in increasing the signal-to-noise ratio still further, there is an additional factor that may limit the maximum obtainable resolution, the molecular diffusion. This averages the spatial information over the dimension of the diffusion of the observed molecule. For pure water the mean displacement during the typical duration of a spin echo, $T_E = 10 - 20$ ms, is at the most two to three micrometers, and under biological conditions it is considerably less.

Two examples are given below of the application of NMR microscopy at 300 MHz (7 tesla) with a gradient of 1.0 mT/cm; the matrix in the image plane consists of 256^2 pixels. Figure 4.29 shows a cross section of a mouse kidney with

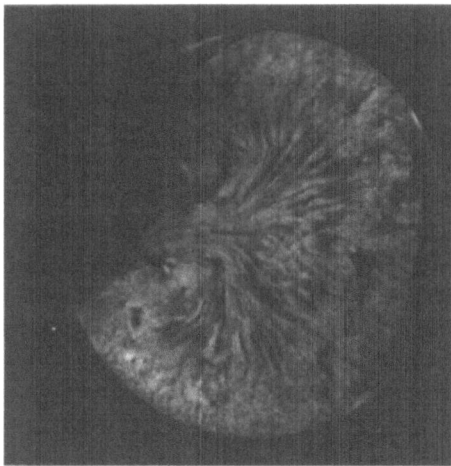

Fig. 4.29. NMR image of a mouse kidney. Cross section of a mouse kindney of about 8 mm diameter. $B_0 = 7\,T$, gradient strength 1 mT/cm, 2D FLASH technique, $T_R = 200\,ms$, pulse angle $\alpha \approx 35°$, $T_E = 8\,ms$, resolution $36^2\,\mu m^2$, slice thickness 320 μm (Bruker, Karlsruhe)

a diameter of about 8 mm recorded with a FLASH pulse sequence. The resolution in the image plane is $36^2\,\mu m^2$, and the slice thickness 320 μm, which is a lower resolution by almost one order of magnitude perpendicular to the image plane. The inner structures can be seen very clearly even with such a small object.

Figure 4.30a shows the cross section of a leaf of *Bryophyllum tubiflore*, with the diameter of the object about 3 mm. The image was made with the spin echo technique, and the resolution is $15^2\,\mu m^2$ in the image plane with a slice thickness of 220 μm. The larger slice thickness does not appreciably reduce the quality of the image since the object has a fiber structure. In Fig. 4.30b an image of the same cross section taken with a light microscope is reproduced for comparison. While a mechanically thin section of the leaf must be cut in order to make an image

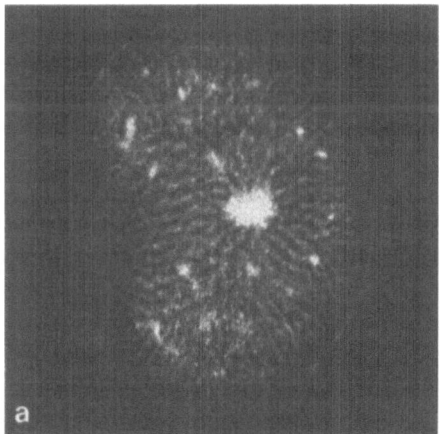
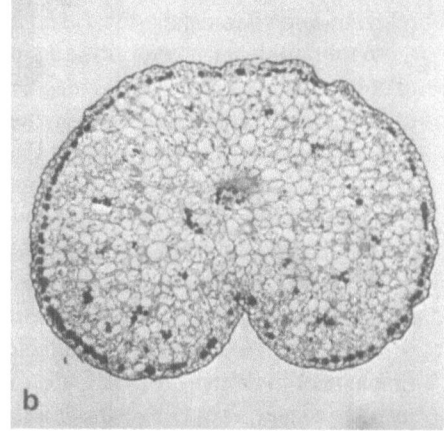

Fig. 4.30a,b. Cross section of a leaf of *Bryophyllum tubiflore*. (a) NMR micro imaging, $B_0 = 7\,T$, gradient strength 1 mT/cm, 2D spin echo technique, $T_R = 1\,s$, $T_E = 16\,ms$, resolution $15^2\,\mu m^2$, slice thickness 220 μm (Bruker, Karlsruhe). (b) Light microscopic image of the same object

with a light microscope, the leaf remains undamaged with the NMR image; the desired plane is selected only by the magnetic field gradient. The bright vascular bundles (dark in the light microscope), through which the water in the leaf is transported, can be easily distinguished.

4.4 Biomedical Effects and Dangers of NMR

An obvious advantage of the application of NMR is that this imaging method does not use ionizing radiation and hence does not present health risks. However, one has to consider a priori side effects, as is always done in the case of physical or pharmaceutical effects on the human organism. The essential question is how serious are these side effects as compared to their medical benefit, and how high is the probability of their occurrence. To assess possible risks, the physical interactions with the human body which play a role in NMR tomography, and can cause biological effects, must be investigated. From the point of view of physics it makes sense to discuss separately the influence of the static magnetic field, of the gradient field and of the radio frequency field.

4.4.1 Biomedical Effects of the Static Magnetic Field

The static magnetic field B_0 has its main effect on paramagnetic and ferromagnetic substances. In the homogeneous magnetic field in the inner part of the magnet the existing magnetic dipoles are aligned, and in addition, in the inhomogeneous magnetic field in the outer parts a strong force is exerted in the direction of the field gradient. If these forces act on macroscopic bodies, they present a serious source of danger. Mobile ferromagnetic objects, such as scissors and scalpels, are attracted very strongly by the magnet and can injure people within their trajectories. These forces can, of course, act also on ferromagnetic implants (pace makers, hemostatic clips) and change their positions.

At the molecular level, in principle the same effects occur if the molecules under consideration possess a magnetic dipole moment. The alignment of lipids in membranes and of suitably structured proteins in high magnetic fields is well known (see Sect. 3.2.5), a method that is used deliberately in biophysics for obtaining oriented structures. This alignment of the molecules can in principle, of course, lead to disturbances of the natural functions. However, the forces acting on the molecules by thermal collision processes, which are normally compensated, are of comparable magnitude; hence, probably very specific processes would have to occur to cause relevant medical disturbances.

Since little experience is available regarding dangerous side-effects, the recommended limits are, depending on the country, in the field range of 2.0 to 2.5 tesla. However, it is to be expected that these limits will be increased in the case of a real clinical need for higher fields.

An indirect source of danger, finally, is mechanical or electrical malfunction of equipment that could occur near the NMR magnets.

4.4.2 Biomedical Effects of Gradient Fields

Time-dependent gradient fields are essential for NMR tomography. Gradient fields are much smaller than B_0-fields, but they are per se strongly inhomogeneous. Hence, the static effects are similar to those of the B_0-fields. In addition, there is the time-dependent component. Time-dependent magnetic fields induce in conductors, and thus also in the human body, induction currents that can lead to depolarizations of the membranes. Beside these electrical effects, time-dependent mechanical deformations can also be produced. They are most likely the cause of the phosphenes which have been known for a long time, which are felt by patients as optical sensations (lightning phenomena), and which may occur when the gradient fields are switched on or off.

The strength of the induction currents is proportional to the time-dependent change of the magnetic field dB/dt; therefore, safety recommendations restrict the maximum change of the magnetic field per unit time. Typical values are between 3 T/s and 20 T/s and these limits depend reasonably on the duration of the change of the magnetic field (usually in the millisecond range).

4.4.3 Biomedical Effects of the Radio Frequency Field

As discussed above, the energy of the radio frequency quanta is too small by several orders of magnitude to break covalent chemical bonds. The main effect of radio frequency radiation that is deliberately used therapeutically with diathermy is the heating of the tissue. The absorption of radio frequency energy is frequency-dependent and increases with increasing frequency. Furthermore, it depends on the composition of the tissue and increases with the ionic strength of the medium.

The human organism produces heat permanently even at rest (approximately 1 W/kg). Hence, an absorption of energy well below this value can clearly be considered harmless. In daily life the physiological mechanism of heat regulation must deal with considerably higher quantities of heat produced by physical activities or resulting from sources outside (e.g. radiation of sunshine, air temperature). Therefore, reasonable limits will be somewhere in the range between the basic rate of heat production and the maximum amount of heat which temperature regulation mechanisms can dispose of without a serious increase in temperature. Typical limits are several W/kg tissue or are directly related to the increase in temperature of the body that should be typically below 1° C.

5. Spatially Selective Spectroscopy and In Vivo NMR

NMR spectroscopy on living objects differs from the high resolution NMR in homogeneous media described in Chaps. 2 and 3 in one significant aspect: living systems have a complex structure in spatial dimensions and hence it is in most cases necessary to use suitable methods to take this spatial dependence into account. The simplest of these systems are cell components and organelles al-Living cells can ultimately form tissues, which then again as a next step of oraganization form whole organs. The investigation of complete higher organisms is ultimately what is understood in medicine as in vivo NMR (In medicine in vivo NMR is often simply called magnetic resonance spectroscopy or MRS). It requires in general a technically more sophisticated spatial selection than the NMR of isolated cells. However, the methods and questions that can be investigated with NMR do not in principle present anything new compared to those with cell components or whole living cells, the only difference being that the experiments are more complex due to the necessary spatial selection.

Living systems consist of many components; for this reason one would expect an NMR spectrum to consist of a superposition of spectra of many different molecules which are hard to analyse and to assign. Fortunately, in practice the situation is frequently more favorable: only a few of the components have such a high concentration that their NMR signals can be distinguished from the background. Hence, the spectra are usually comparatively simple and can be interpreted without undue effort.

In order to obtain additional spectroscopic information besides the spatial information, the signal intensity is distributed over an additional dimension. For this reason problems with the detection sensitivity often arise. If the spatial selection is reduced by averaging over larger volumes, metabolites with lower concentration can be detected, and vice versa. Consequently, one must always compromise between spatial selection and detection limitations (except with methods that utilize the internal properties of biological systems for a rough spatial selection). While the discussion of the ideal field strength in the case of pure tomography is still continuing, for spectroscopic purposes the highest possible magnetic fields must be applied because spectral resolution and sensitivity both increase with increasing field strength. However, it is technically difficult to build magnets with very high field strengths and large inner diameters. While at present it is possible to obtain commercial magnets with a field strength of 14 tesla (600 MHz proton resonance) for high resolution NMR with sample diameters of 10 mm, the maximum field strength that has been obtained so far with a diameter of 1.2 m is 4 tesla (170 MHz proton resonance).

5.1 Methods for Spatial Selection

In in vivo NMR there are two qualitatively different types of spatial information. One is the information on the absolute position in the laboratory system that we obtain typically for water molecules with NMR tomography. The other is the information on the relative position with respect to certain biological structures, that is, whether the molecules investigated are inside or outside of defined closed structures, such as cells or compartments.

All methods of spatially selective NMR spectroscopy require spatial averaging to obtain a sufficient signal-to-noise ratio. In general, this averaging comprises a large number of cells and compartments. For this reason one must always be very cautious when interpreting the spectroscopic data on the morphological level. This is particularly critical if one wants to establish the functional state of individual cells in heterogeneous tissue on the basis of measurements of the substrate concentration. However, even in suspensions of similar cells one must take into account the fact that averaging is done over many different regions in which totally different conditions may exist.

The main problem with spatially selective in vivo spectroscopy is the signal-to-noise ratio. Besides the lower B_0-field strength of the large-volume whole-body magnets mentioned above this is mainly due to the filling factor. The filling factor is defined as the ratio of the volume of the sample investigated to the volume of the detection coil, and it should be as close as possible to one. It is easy to calculate that with a measured volume element of $1 \, cm^3$ in a whole body coil with a diameter and a length of about 60 cm the filling factor is of the order of 10^{-5}.

Two different methods were applied to achieve the aim of a spatially selective in vivo spectroscopy: either the spatial selection occurs by spatially selective radio frequency fields that are usually produced by surface coils, or a defined volume is selected similar to NMR tomography by superposing three magnetic field gradients perpendicular to each other on the static B_0-field. The two methods can also be combined, for instance, by applying one gradient along the axis of the surface coil. Another possibility is to excite with a whole-body coil in combination with a suitable gradient and to detect the signal with a surface coil. Surface coils have the great advantage over whole-body coils that the filling factor is better by several orders of magnitude. However, they have the disadvantage that they can only be used near the surface of the body and are therefore inappropriate for investigating organs inside the body. It remains to be seen whether in the future coils will be applied endoscopically in special cases.

5.1.1 Spatial Resolution with Isolated Cells and Organelles

The signals of molecules in two different regions can most easily be distinguished if the corresponding spectra of at least one of these regions can be measured separately. For example, in the case of cell suspensions the cells can be centrifuged and the spectra of the extracellular space (supernatant) and the intracellular space (precipitate) can be measured separately. In general, with in vivo experiments the medium has to be continuously exchanged by a suitable apparatus to keep the cells alive. This provides the possibility of determining the substrate concentration in this solution separately.

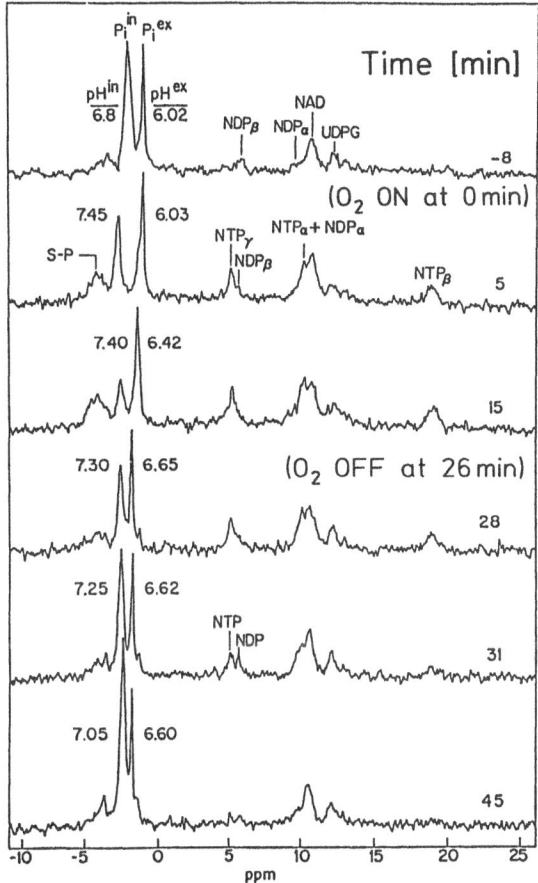

Fig. 5.1. ^{31}P-NMR spectra (146 MHz) of *E. coli* cells under various conditions. The upper spectrum is the starting spectrum after inserting the bacteria into the sample tube. Subsequently, the cells were at first provided with oxygen which was later removed. The last spectra were recorded under anaerobic conditions after the indicated times. P_i^{ext}, external inorganic phosphate, P_i^{int}, internal inorganic phosphate, NDP and NTP, nucleotide di- and triphosphates (α, β and γ denote the corresponding phosphate groups), S-P, sugar phosphate and UDPG, uridine diphosphate glucose (Ugurbil et al., 1982, with permission)

It is certainly more elegant if the signals of the extracellular and intracellular spaces can be observed with one single experiment. This is always possible if the two environments have different influences on observable spectral properties. One of these differences is a possible variation in the susceptibility of the medium in the inner and outer space that causes differences in the chemical shift. In addition, these susceptibility differences may cause strong local inhomogeneities of the magnetic field in the transitional region resulting in different relaxation rates.

The intracellular and extracellular spaces differ frequently in their pH values and ionic composition, which may both have a marked influence on the chemical shifts of the substances contained. These differences of chemical shifts of a substance are especially large if the pK value of a titratable group is close to the pH value. In particular the phosphorus resonances of the frequently occurring phosphate groups show large variations of their chemical shifts if the pH value of the medium is changed, as was already demonstrated for inorganic phosphate in Sect. 2.2.3 (Fig. 2.8). Figure 5.1 shows as an example the ^{31}P-NMR spectrum of a suspension of $E.\ coli$ cells under different conditions. The signal from the external inorganic phosphate can be clearly distinguished from that of the internal inorganic phosphate.

Another method for distinguishing between the inner and the outer space consists in the addition of paramagnetics to one of the two regions. Depending on the paramagnetic used, a broadening or an additional shift is found in the resonance lines of those molecules that come into contact with the paramagnetic center. Any conclusion regarding the localization of the molecules presupposes, however, that the paramagnetic agent remains essentially concentrated in one of the regions, that is, the membrane is barely permeable for it. We have already seen examples of this application: the distinction of lipids on the inner side of the membrane from those in contact with the outer medium (Fig. 3.47) and the distinction of tissues that are separated by the blood-brain barrier (Sect. 4.3.3).

5.1.2 Spatial Selection with Space Dependent Radio Frequency Fields

In every ordinary NMR experiment where the sample extends beyond the radio frequency coil, in a certain sense a space selective experiment is performed because only signals from inside the coil are detected. There are two reasons for this: (1) The radio frequency excitation outside the coil is low, and (2) the detection sensitivity for signals from the outer region is also strongly reduced.

The radio frequency field B_1 is homogeneous only inside the coil and decreases rapidly in the outer region. The spins outside the transmitter coil are therefore insufficiently excited by the radio frequency pulse and for this reason contribute very little to the total signal. The detection of NMR occurs by the induction voltage that is induced in the coil by the magnetic field of the precessing nuclear spins. The larger this induction voltage, the more sensitively can the spins be detected. The spatial dependence of the sensitivity can be easily estimated by using the principle of reciprocity: the induction voltage that the time-dependent magnetization produces at a point P in space in the coil is proportional to the B_1-field which the coil would produce at point P if a current of corresponding frequency were flowing through the coil. Since the B_1 field is normally highest inside the coil and decreases rapidly outside it with increasing distances, the sensitivity of detection for spins is highest in the central region and decreases rapidly beyond it according to the reciprocity principle.

With the first experiments on living animals the spatial selection was in fact achieved with the normal receiver coil of a high resolution spectrometer by placing the part of the body of the animal being investigated in the central region of the receiver coil where it was examined spectroscopically. The next important step was to deliberately choose a configuration of the coil with a strongly inhomogeneous field distribution, namely the surface coil (Fig. 5.2). The majority of in vivo experiments with large living objects are performed even today using this simple set-up.

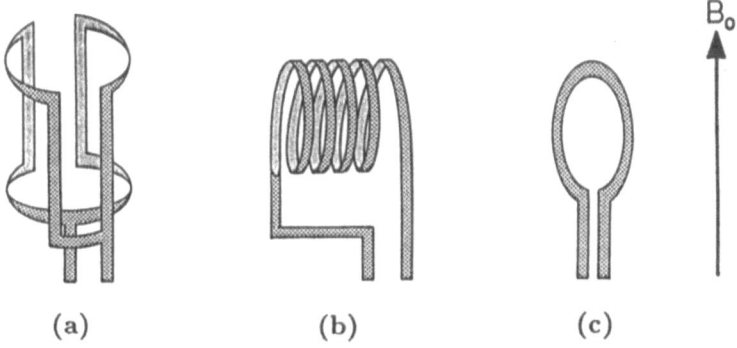

(a)   (b)   (c)

Fig. 5.2a–c. Simple receiver systems. (a) saddle coil (modified Helmholtz coil), (b) solenoid, (c) surface coil. The direction of the external applied field B_0 is indicated

If a flat, circular surface coil is placed on the surface of an object and is tuned with a corresponding capacitance to the resonance frequency of the nuclear spins to be investigated in the magnetic field B_0, then one obtains without additional precautions only signals from the volume elements that are close to the measuring coil (Fig. 5.3). Here the contribution to the signal of volume elements immediately below the surface is largest and decreases continuously along the axis of the coil perpendicular to the surface. Different ways were chosen to further limit the volume investigated. One possibility arises from the fact that the radio frequency field B_1 is always very inhomogeneous over the volume of the sample. The limitation to a volume immediately under the surface is comparatively simple to achieve because for geometrical reasons this volume contributes most of the signal anyway. Let us assume that we choose a pulse sequence which starts with a 90° pulse; if the B_1 field is set in such a manner that it corresponds to a 135° pulse in the center of the coil, it decreases along the axis of the coil and reaches a

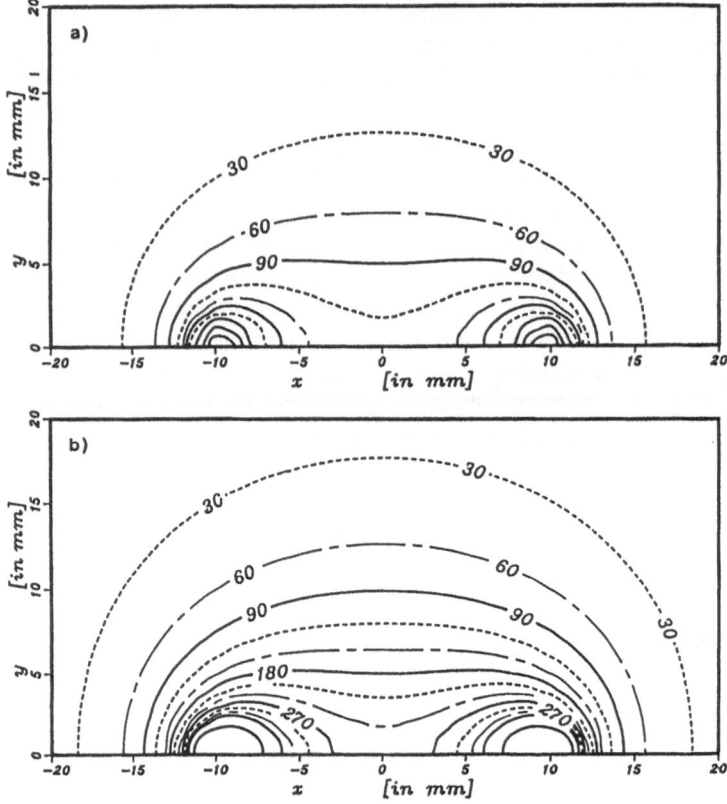

Fig. 5.3a,b. Flip angle distribution with surface coils. The lines of constant flip angle were calculated for a plane perpendicular through the center of a surface coil. The surface coil with the radius $R = 10\,\text{mm}$ is located in the x, z plane. The pulse lengths were chosen so that with $(x, y) = (0, r/2)$ the flip angle $\alpha = 90°$ (a) and $180°$ (b), respectively (Haase et al., 1984, with permission)

flip angle of 45° at a distance from the surface equal to one radius of the coil; the contribution from more distant volume elements is negligibly small. The region from which signals can be detected with sufficient sensitivity corresponds in this simple case to a hemisphere having the same radius R as the surface coil and its center at the center of the coil. However, if one wants to apply the same procedure to a region located deeper inside the body, for example by making the B_1 field so strong that it causes a 90° ± 45° pulse at a distance of 0.7 to 1.5 R, difficulties are encountered. Actually, closely below the surface such a strong B_1 field causes a flip angle of 270° ± 45° that has the same effect as the 90° pulse, that is, to flip the magnetization into the x, y plane. Since the signal from this region, in which a certain radio frequency current produces a three times higher B_1 field, also induces a three times higher voltage in the radio frequency coil because of the reciprocity principle, it contributes more to the total signal than the volume desired.

When applying surface coils one performs a double spatial selection, one spatial selection with the excitation and one spatial selection with the detection. However, both functions can be separated by a suitable set-up. For certain spectroscopic experiments a homogeneous B_1 field for excitation is required. This can be produced with a whole body coil and the signal can nevertheless be detected with an additional surface coil.

The spatial selectivity of the excitation can be increased by spatial pulse sequences that are based on the spatial dependence of the flip angle. Remaining signals from unwanted regions are usually eliminated by special phase cycles, which cause a cancellation of the signals when adding up the spectra of these regions, while the desired signals remain. Many pulse sequences have been suggested in order to reach this goal; the most generally known are the depth pulses. The simplest of these pulse sequences consists of a pulse of length t_α, immediately followed by a pulse of length $2t_\alpha$, the phase of which is cyclically varied in 90° steps. At positions at which the pulse length t_α corresponds to a 90° pulse, the second 180° pulse just leads to an inversion, while in other regions the signal intensity is reduced (Bendall and Pegg, 1985).

A shortcoming of all these methods is that the volume chosen has an irregular shape, so that an exact selection of the spatial position is very difficult. For structures near the surface, however, the detection with surface coils has in any case the great advantage of a higher sensitivity because of the much better filling factor. This higher sensitivity is sometimes even used with tomography for imaging structures near the surface when surface coils are used for the detection of the signals.

5.1.3 Spatial Selection with a Space Dependent Static Magnetic Field B_0

Spatial selection based on the spatial dependence of the B_0-field by superposing it with magnetic field gradients requires a higher technical effort than the methods described above, but it can be more generally applied. A basic problem that remains is the signal-to-noise ratio. Two types of methods can be distinguished: localization techniques that produce highly resolved spectra similar to spectroscopy from a single volume element as well defined as possible, and tomographic methods where spectroscopic data are collected simultaneously from a number of volume elements. The localization techniques correspond to the sequential point methods used in the early days of NMR tomography that are today completely outdated because they constitute a very lengthy method for composing a complete image. With spectroscopy the situation is different since in many cases one is only interested in the spectra of a particular region, where for instance a tumor is located, while information on other regions are irrelevant.

Most tomographic techniques can be slightly modified in such a manner that additional information on the chemical shift is included. To keep the line broadening low and thus obtain spectral information in high resolution, the magnetic field must be as homogeneous as possible when recording the spectral information. Therefore, the gradients have to be turned off during the recording time. Even switching off gradients immediately before the start of recording affects the homogeneity so strongly, by inducing eddy currents in the metal of the probe of the magnetic field, that this switching should be completed some time before the recording (about 10 ms).

An ideal technique would meet an additional requirement which has so far not been met by any method: the spectroscopic information should not be affected by coding the spatial information and it should be possible to apply all the experimental schemes known from high resolution NMR spectroscopy without impairment.

Of the many possible techniques in this area, we wish to describe just one, the LOCUS (localization of unaffected spins) spectroscopy (Fig. 5.4). With this technique the selection process leaves the spins of the selected region essentially unaffected, so that subsequently in the absence of gradients any NMR experiment can be performed in this region. For this purpose the gradients are switched on and the magnetization of the surrounding regions is flipped in the x, y plane by frequency-selective 90° pulses that do not contain resonance frequencies of the region selected; it dephases fast in the gradient fields. The efficiency of this procedure can easily be checked by recording a normal tomogram after the selection procedure, which in the ideal case should only show the region selected. The selectivity of this method is limited by the longitudinal relaxation of the surrounding spins that cancels the saturation effect.

As compared to localization techniques, true spectroscopic imaging has the advantage that the region of interest can be selected after performing the experiment, and at least in principle a spectrum can be obtained of any anatomical region of interest, however complex its shape. Figure 5.5 shows a modification of the spin echo experiment known from tomography (Fig. 4.4), which is suit-

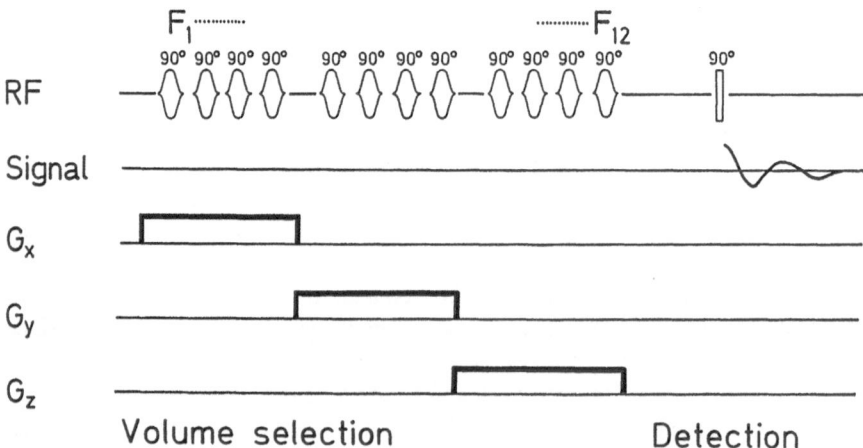

Fig. 5.4. Schematic plot of the LOCUS experiment. $F_1, \ldots, F_{12}$ are frequency selective 90° pulses with different carrier frequencies that act on the sample simultaneously with the gradients G_x, G_y and G_z. The signal obtained after a final 90° pulse is shown (Haase, 1986, with permission)

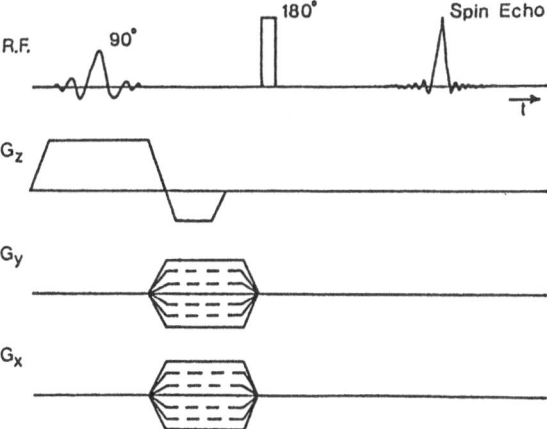

Fig. 5.5. Schematic plot of a spin echo experiment for spectroscopic imaging. In this 2D version of the spin echo experiment the G_z gradient defines a plane together with the selective 90° pulse. The strengths of the gradients G_x and G_y are varied independently of each other from experiment to experiment (Maudsley and Hilal, 1985, with permission)

able for obtaining spectroscopic information from a preselected plane. The only difference with respect to pure imaging is that the constant reading gradient in the x direction, which is otherwise switched on when recording the signal, is in this case applied during the evolution time. For spatial coding the x and the y gradients must now be varied step by step from experiment to experiment. The corresponding 3D experiment can be obtained by replacing the selective 90° pulse by a non-selective one and by changing the z gradient as well, step by step like the other gradients. However, in this case the minimum number of individual experiments increases considerably, and with it the minimum duration for the 3D image.

5.2 Typical Applications and Problems
of NMR Spectroscopy in Living Systems

In this section we shall show biochemical information can in principle be obtained with in vivo NMR. When choosing the examples no particular emphasis will be placed on the method by which the spatial selection was performed; with most examples, either the spatial selection is unnecessary because of the set-up of the experiment or it is carried out by the simple application of surface coils. This reflects the state of the art at present, where one is still faced with so many spectroscopic and technical problems that whenever possible one tries to avoid additional problems which arise from complex spatial selection methods with large living objects, such as the human body. Nevertheless, these examples show the potential possibilities which exist for investigations of human subjects in the future.

A frequent criticism of in vivo NMR is that is has so far almost exclusively merely confirmed already known findings which had been obtained previously with other methods. This is certainly true, but it is not an argument against in vivo NMR as such, but rather for the ingenuity of biochemists who have obtained the right results with the classical biochemical methods. Nevertheless, in vivo NMR offers something new, which to this degree cannot be achieved by any other method, namely the non-invasive and continuous observation of cellular metabolic processes on the molecular level. Furthermore, in many cases it permits to directly measure quantities that can be obtained only in a very complicated way with other methods.

5.2.1 Preservation of the Natural Functions

An important precondition for meaningful in vivo spectroscopy is that the object to be investigated is in a well defined and readily reproducible state. In most cases this state should be as close as possible to the intact physiological state. Biological systems are in general not in equilibrium, but the homoeostasis can only be maintained by changes of the total system consisting of the biological system and its environment: these changes have to be taken into account, especially for long-lasting experiments.

Accordingly, there are several solutions adapted to the various special problems. If one is concerned with simple particulate systems, where only substrates and products in the outer medium are to be observed, it is sufficient to prevent the sedimentation of the corpuscular elements and to use substrates in large excess. Frequently it suffices to rotate the sample, eventually combining this with a stirring mechanism to prevent sedimentation. It is even simpler when studying the internal metabolism of cell and tissue samples of low metabolic activity. Here the cells can be packed densely or the tissue fixed in suitable sample tubes. In general, however, one wants to control the concentrations of the substrates and the metabolites in the outer medium by exchanging the outer medium in a well defined manner. Figure 5.6 shows a typical set-up. An excellent way to prevent sedimentation of the cells and to assure the exchange with the surrounding solution is to embed the cells into thin agar threads, thus preventing the cells from flowing out of the sample volume. Depending on the experimental requirements numerous other set-ups can be applied; for instance it is possible to perform the exchange of matter through semipermeable dialysis fibers or to fasten the cells on rigid carriers with large surfaces. With larger cellular units there is no danger that they flow out with the perfusion solution. However, there is the problem that with too-large tissue sections long diffusion paths may hinder the effective exchange of matter. Under favorable conditions it is possible to grow so-called spheroids from single cells, that is, small heaps of cells small enough (approximately $100\,\mu$m) not to hinder markedly the diffusion, but big enough to be prevented from flowing out. Ultimately, with whole organs there remains only the perfusion of the organs with suitable oxygenated media. This problem

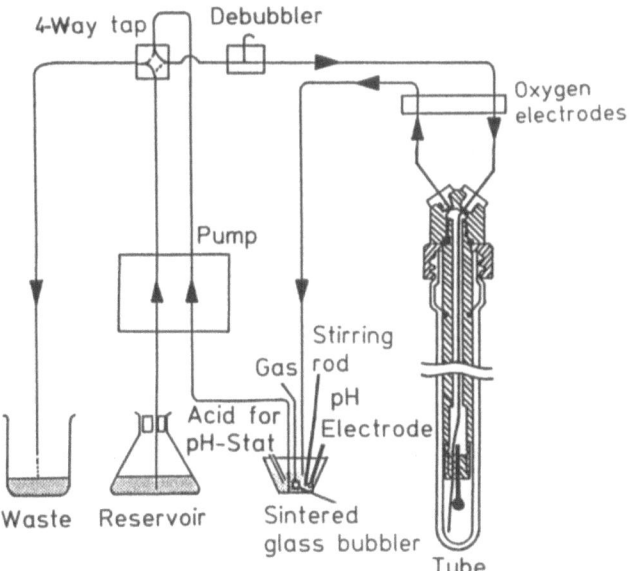

Fig. 5.6. Perfusion apparatus for in vivo NMR experiments with isolated cells and tissues. In the set-up shown the partial oxygen pressure and pH value of the medium are regulated; if necessary, the medium itself can be renewed. The temperature is regulated by the temperature control unit, which is an integral component of every high resolution spectrometer, but can in addition be stabilized by an external temperature control of the medium (Lyon et al., 1986, with permission)

is, however, not new in biochemistry and physiology. The existing methods just have to be adapted to the spatial restrictions within an NMR spectrometer.

Although during experiments with animals the conservation of the physiological functions is to a large extent maintained automatically, problems may arise because experiments are in general done with animals under anaesthesia. To ensure that reliable results are obtained, the vital functions have to be carefully controlled. NMR spectroscopy with human beings is in general not dangerous (see Sect. 4.4); nevertheless, stricter precautions have to be taken when carrying out these investigations.

5.2.2 In Vivo ^{1}H-NMR Spectroscopy

At first glance ^{1}H-NMR appears to be especially suitable for in vivo applications because it has a very large detection sensitivity, and the high natural abundance of the ^{1}H nuclei makes isotopic labelling unnecessary. On the other hand, the comparatively small range of chemical shifts over which the resonance frequencies of biological molecules extends (see Sects. 1.2.1 and 2.11) impedes the resolution of individual resonances. Furthermore, the intense water signal disturbs the observation of the signals of the particularly interesting metabolites. Frequently the water signals can be sufficiently weakened by selective excitation or selective saturation (see Sect. 2.1.7).

A strong superposition of resonance lines is very often observed with in vivo NMR spectra. Here one can try to obtain the information desired by a deliberate simplification of the spectra, a process that is frequently termed spectral editing. A very simple possibility that we already know is the spin echo sequence $((\pi/2)_x - \tau - \pi_y - \tau)$. It leads to a T_2 weighting, and in addition it changes the phases of the signals depending on the multiplet structure. Hence, spin echo spectra contain not only positive signals, but also signals that have negative components. This phenomenon can easily be described with the classical vector model (Fig. 5.7). The 180° pulse does indeed lead to a refocusing of effects owing to field inhomogeneity or chemical shifts, but the dephasing of the individual components of the multiplet due to J coupling remains unaffected. After a time 2τ, which equals $1/J$, the components of a doublet point in the $-x$ direction and thus give a negative signal, while singlet resonances or the central component of a triplet point in the $+x$ axis and therefore give a positive signal.

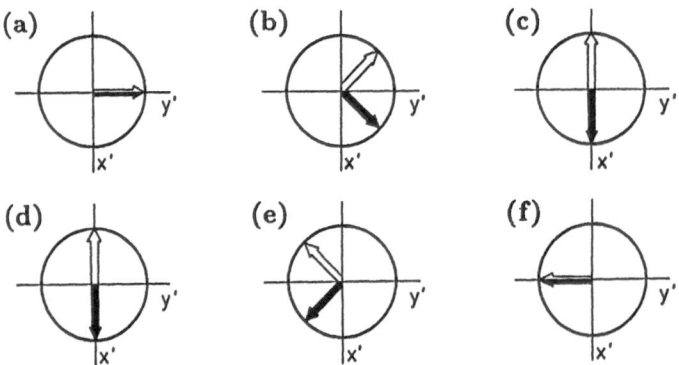

Fig. 5.7a–f. The spin echo experiment with J coupled systems. Schematic plot of the magnetization components of a doublet resonance with a pulse sequence $(\pi/2)_x - \tau - \pi_y - \tau$ in the rotating coordinate system. Magnetization after the $(\pi/2)_x$ pulse (a), at time $\tau/2$ (b), at time $\tau = 1/2J$ before (c) and after (d) the π_y pulse, at time $3/2\tau$ (e) and at time $2\tau = 1/J$ (f). At (a) both doublet components are in phase, at (c) and (d) in antiphase and at (f) again in phase, however, with an inverted sign with respect to (a)

Figure 5.8 shows spin echo spectra of red blood cells, which are very easy to investigate because their metabolic activities are rather low and anaerobic. Hence, they can be packed into a sample tube in a suitable buffer as densely as possible and easily examined spectroscopically without great precautions. In the spin echo spectrum the broad signals of macromolecular components are suppressed to a large extent; because of their short transverse relaxation time T_2, at the time of the echo their magnetization has almost decayed nevertheless, the signals of a number of different molecules are observed. Although the signals of small molecules are strongly enhanced by this method, parts of the signal of hemoglobin still remain because it occurs in erythrocytes in very high concentration. If one proceeds carefully and takes the T_2 weighting into account, this method permits one to

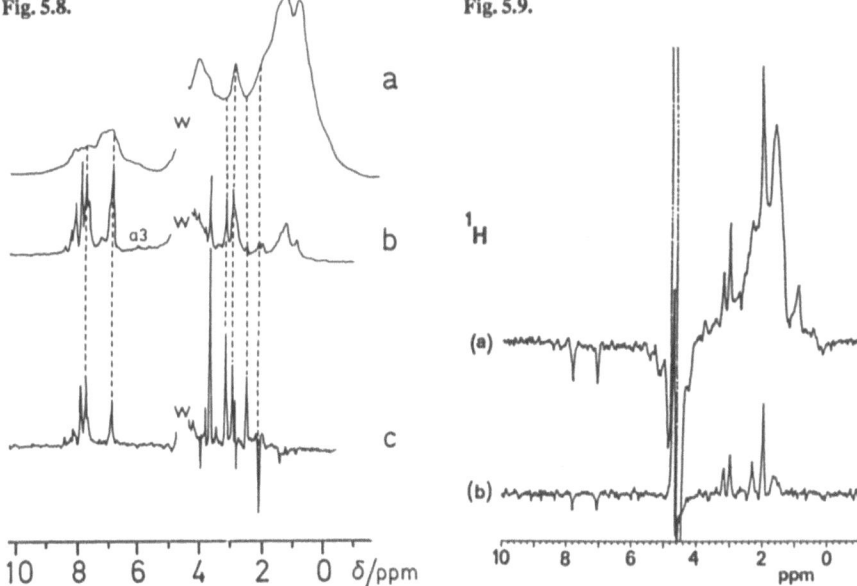

Fig. 5.8. ^{1}H-NMR spin echo spectra of red blood cells (270 MHz). Normal Fourier NMR spectrum (a), spin echo spectrum with a τ of 20 ms (b) and of 60 ms (c). In the low field range some signals are observed from the aromatic rings of the nucleotides and the histidine residues of the hemoglobin. In the high field range the signals are from glutathione, pyruvate and lactate (Brown et al., 1977, with permission)

Fig. 5.9. ^{1}H-NMR spin echo spectra (360 MHz) of the brain of histidinemic mice recorded with a surface coil. Delay time τ of 20 ms (a) and 60 ms (b). The signals of the ring protons of histidine (7.83 ppm and 7.08 ppm), of the methyl protons of choline (3.20 ppm), of creatine and phosphocreatine (3.03 ppm) and of N-acetylaspartate (2.01 ppm) can be clearly recognized (Gadian et al., 1986, with permission)

determine the intracellular concentration of the metabolites and to follow their changes as a function of the experimental conditions.

Histidinemia is an innate defect of metabolism involving a lack of histidase, an enzyme that decomposes the amino acid histidine. This explains why higher concentrations of histidine are found in blood and tissue. With human beings this defect is accompanied by a disturbed mental development. Considering an animal model, the histidinemic mouse, the resonances of histidine can be clearly detected in the brain tissue with the spin echo method. Figure 5.9 shows such spectra recorded with a living mouse using a surface coil. Beside the resonances of the histidine protons at 7.83 ppm and 7.08 ppm one recognizes among others the intense signals of the methyl protons of choline, creatine, phosphocreatine and N-acetylaspartic acid.

Numerous methods have been developed for eliminating from the spectrum certain multiplets which are of no interest. For instance, if one adds a spin echo spectrum, in which the doublet components with the coupling J are just in antiphase, to a normal spectrum, one obtains a spectrum from which these doublet resonances have vanished to a large extent.

With a variation of this difference method resonances can be selected that couple with a certain nuclear spin. In this case two differently produced signals are subtracted from each other, a normal spin echo spectrum where τ is again chosen in such a way that the signal of interest is just negative, and a spin echo spectrum where one decouples frequency selectively with high power during time τ at the resonance frequency of the coupling partner. The result of this decoupling is that the resonance lines of the corresponding coupling partners, in contrast to the other resonance lines, do not participate in the otherwise observed change of sign. Hence, when forming the difference only this signal remains. Figure 5.10 shows an example of how well this method functions even with biological tissues. In the working muscle lactic acid (CH_3-CHOH-COOH) is produced in high concentrations if the supply of oxygen does not suffice. The methyl signal of lactic acid at 1.33 ppm can be selectively detected with the method described here if the CH resonance at 4.1 ppm is decoupled. It is remarkable that this detection still functions even if the multiplets are unresolved because of the field inhomogeneity.

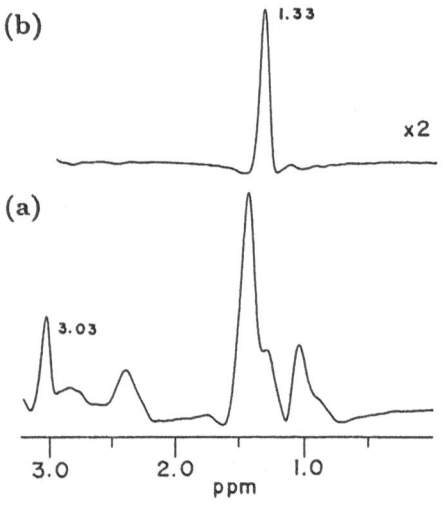

Fig. 5.10a,b. ^{1}H-NMR spin echo double resonance spectra of a rat muscle at 360 MHz. (a) Section of a normal spin echo spectrum with a delay time τ of 68 ms, (b) spin echo double resonance spectrum. The latter was made while recording every second FID, with the resonance at 4.1 ppm decoupled and the corresponding signal subtracted from the spin echo signal without decoupling (Rothman et al., 1984, with permission)

5.2.3 In Vivo ^{31}P-NMR Spectroscopy

Typical properties of ^{31}P-NMR are a large range of chemical shifts, which for biologically important molecules is around 30 ppm, a not-too-low NMR sensitivity of about 7 % of the proton sensitivity and a natural abundance of 100 %. Furthermore, problems due to the superpositions of resonance lines are small compared to proton resonance because with most investigations only a few different molecules containing phosphorus are present. As a result of these properties

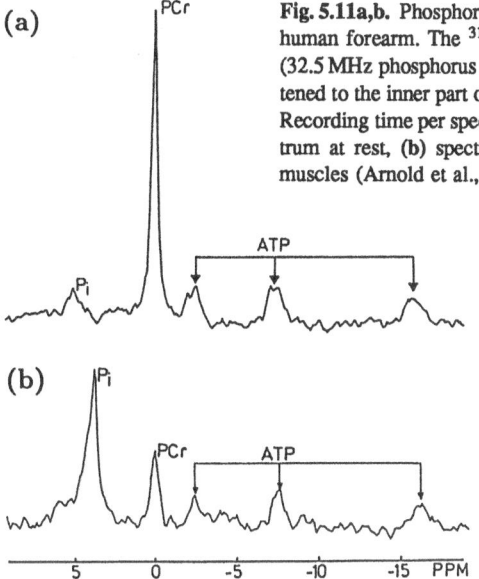

(a)

(b)

Fig. 5.11a,b. Phosphorus resonance spectrum of the muscles of a human forearm. The ^{31}P-NMR spectra were recorded at 1.89 tesla (32.5 MHz phosphorus resonance frequency) with a surface coil fastened to the inner part of the forearm (flexor digitorum superficialis). Recording time per spectrum 64 s (a), 32 s (b) respectively. (a) Spectrum at rest, (b) spectrum after 2.5 min of exercise of the flexor muscles (Arnold et al., 1984, with permission)

a large part of all in vivo applications of NMR are experiments with phosphorus resonance.

In biologically active systems, depending on the functional state, a number of different phosphorus-containing metabolites can be detected with ^{31}P-NMR. However, if their concentration is below 0.1 mM, they are in general below the limit of detection. Because of the physiological necessity of a sufficiently high intracellular ATP level, the three phosphorus resonances of the phosphate groups of ATP can almost always be observed in living cells. In addition, in most cases the resonance lines of inorganic phosphate P_i and phosphocreatine are found.

All these signals are usually detected in spectra of muscles. Figure 5.11 shows as an example phosphorus resonance spectra that were taken with a surface coil from the flexor muscles of a human forearm. Increased muscular effort leads to a decrease of the phosphocreatine signals and an increase of the P_i signals. This means that under these conditions the concentration of the phosphocreatine decreases, the concentration of the inorganic phosphate increases and the ATP concentration remains unchanged. This is a reasonable result: the hydrolysis of ATP to ADP and P_i produces the energy needed for the muscle contraction. In order this process can continue, ATP must be constantly resynthesized and the ATP level kept steady. The resynthesis occurs by several pathways. Substrates such as glucose are fed into glycolysis and into oxidative phosphorylation to form ATP in the same way as occurs to a lesser extent at rest. When ATP consumption is rapid, ATP is also regenerated from the phosphocreatine stored at rest, the phosphate group of which is transferred to ADP. Hence, under these conditions the phosphocreatine level decreases and the phosphate level increases.

As described previously (Sect. 2.2.3), the phosphate group occurs in different forms that interconvert in solution by protonation/deprotonation processes and are in equilibrium with each other. Usually the condition of fast exchange is satisfied for the transition between these forms and one observes a single resonance line that is shifted depending on the pH value of the solution. This is also the reason why the resonance lines of the inorganic phosphate appear at different positions in the two spectra of the arm muscles in Fig. 5.11. Obviously, the pH value in the muscle has changed. The mathematical relation is rather well described by the modified Henderson-Hasselbalch equation (2.16 and 2.17). If the pK values of the equilibrium between the different forms and the corresponding chemical shifts are known, the chemical shift for each pH value can be calculated using (2.17); conversely by measuring the chemical shift the corresponding pH value can be determined.

The inorganic phosphate can thus be used as a natural pH indicator. With pH values close to the apparent pK value pK' of about 6.9, small changes of pH cause large changes in the chemical shift. For this reason, changes in the pH value in this range can be observed with high precision by measuring the chemical shift of the phosphorus resonance, and pH changes of about 0.02 units can be detected. If one wants to use the pH dependence of the chemical shift for an exact determination of the pH values, one has to consider that in principle all equilibrium constants, and hence the pK values of the phosphate groups also, are temperature-dependent. Furthermore, the ionic strength has an effect on the ionization equilibrium, and specific interactions such as the binding of magnesium and calcium ions can cause changes in the titration behaviour.

For practical applications where a high absolute precision is needed it is therefore advisable always to measure calibration curves with external conditions as similar as possible. An additional possibility of checking exists if the chemical shifts of other ionizable phosphate residues are observed simultaneously. The reproducibility of the measurements can be considerably improved by using an internal standard. A natural internal standard is the phosphocreatine that is found in many cells and is frequently used as a reference substance. If one makes the simplifying assumption that in the pK range just one pK value suffices for describing the functional connections, the pH value can be expressed as a function of the chemical shift δ of the inorganic phosphate by transforming (2.16):

$$\mathrm{pH} = pK' + \log \frac{\delta - \delta_1}{\delta_2 - \delta} \quad . \tag{5.1}$$

The value of the constants pK', δ_1 and δ_2 varies from tissue to tissue, and the values for the same tissue quoted in literature vary from author to author. For practical purposes this is really not as serious as it may appear because the interesting changes of the pH value are usually predicted comparatively well. Typical values for the skeletal muscle of warm blooded animals are a pK' value of 6.75, a δ_1 of 3.27 ppm and a δ_2 of 5.69 ppm (Taylor et al., 1986).

The pH dependence of the chemical shift has its disadvantages as well. In an unknown mixture of substances it is in general not possible to infer from the chemical shift that a certain substance is present without a precise knowledge of the conditions of measurement. In particular, the uncritical use of values from the literature as standards for comparison can for this reason lead to erroneous interpretations. Furthermore, the application of external references as is usual with phosphorus resonance can cause considerable deviations of the values measured, depending on the geometrical set-up and the susceptibility of the samples (see Sect. 2.1.2); consequently, the use of an internal reference is advisable here also, if possible.

The complex interactions of the enzymes control the metabolic processes in living cells. The enzymatic activities are regulated in many ways: the concentration of the enzymes can be changed at the level of synthesis or breakdown; covalent modification and interactions with other components of the system, with specific regulators, with substrates or with products can determine the actual activity of the individual enzyme in multiple interconnected regulatory networks. It is, therefore, not to be expected that knowledge of enzyme characteristics from in vitro experiments suffices to describe processes that take place in the intact cell. Although substances are continuously synthesized and broken down in the cell – in the human body, for instance, approximately 70 kg of ATP per day are chemically processed – on the whole a relatively stable equilibrium is still sustained. Hence, it is important to have methods at hand that can determine the enzyme activities in the intact cell under these conditions. One possibility for doing this is given by the transfer of nuclear spin polarization mentioned above (see Sect. 2.2.5), which can be determined with the method of selective saturation transfer, with selective inversion transfer or with two-dimensional exchange spectroscopy.

One reaction studied in many tissues with this method is the creatine kinase reaction, which facilitates a fast regeneration of ATP in the case of high ATP consumption from phosphocreatine and catalyzes the resynthesis of phosphocreatine (PCr) from creatine (Cr) when the ATP consumption is low. Both reactions can be described as

$$PCr^{2-} + Mg^{2+}.ADP^{3-} + H^+ \underset{k_{-1}}{\overset{k_1}{\rightleftharpoons}} Cr + Mg^{2+}.ATP^{4-} \quad . \tag{5.2}$$

Thus the phosphate group changes back and forth between two different chemical surroundings, the phosphocreatine and the ATP. Figure 5.12 shows inversion transfer spectra taken with the perfused heart of a rat at different waiting times τ. If the intensity of the resonance lines of phosphocreatine and the γ-phosphate of ATP is plotted as a function of τ (Fig. 5.13), the effect of the polarization transfer becomes still clearer. The experimental results can be described by the equations given in Sect. 2.2.5. The rate constants for the reactions in the heart muscle can then be determined from the parameters used in these equations. A general problem that also affects these experiments is the model dependence of all the results obtained. Cells are not homogeneous but are divided into compartments. The various metabolic processes within the individual compartments occur more or less coupled with each other, but NMR averages over all compartments.

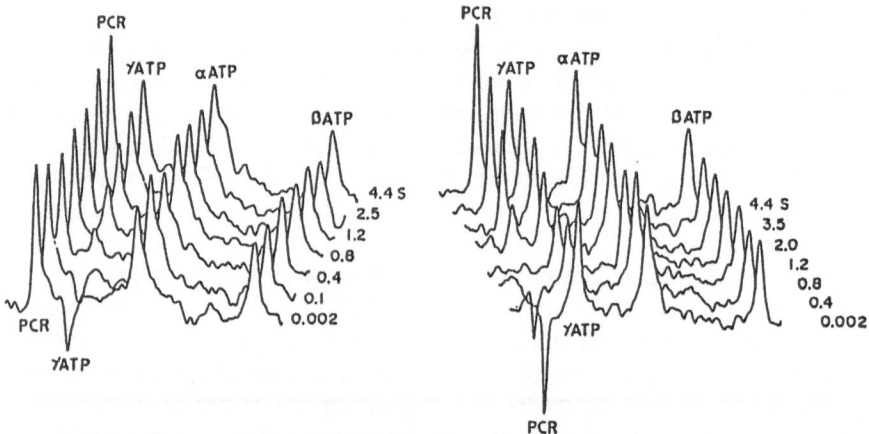

Fig. 5.12. Inversion transfer experiment on the activity of the creatine kinase in the perfused heart of a rat. The ^{31}P-NMR spectra were recorded at 145 MHz (8.5 tesla) at 37°C. Selective inversion of the resonance of the γ-phosphate of ATP (left) and of the resonance of the phosphocreatine (right). The times mentioned refer to the delay time τ between the selective 180° pulse and the 90° pulse (Degani et al., 1985, with permission)

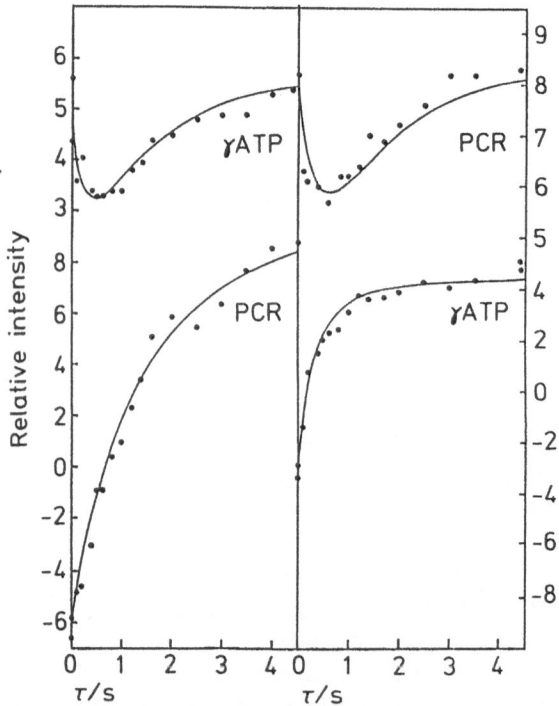

Fig. 5.13. Change of the signal intensities with the inversion transfer in the heart muscle. The values plotted originate from the experiments shown in Fig. 5.12 to which the theoretical course of the inversion transfer was fitted (Degani et al., 1985, with permission). Here the left part of Fig. 5.13 corresponds to the spectra in the right part of Fig. 5.12 and vice versa

190

ATP is produced largely in the mitochondria, while it is consumed mainly in the cytosol for specific functions such as muscle contraction. Phosphocreatine and creatine kinase are found both in the cytosol and in the mitochondria; phosphocreatine can pass through the mitochondrial membrane via a specific transport system, the phosphocreatine shuttle. The rate constants in (2.18) describe the individual processes correctly only if a rapid equilibrium exists between cytosol and compartment.

Almost all organs have been investigated with ^{31}P-NMR. It is comparatively easy to record a ^{31}P spectrum and to assign the resonance lines obtained. The future usefulness of ^{31}P-NMR for in vivo spectroscopy will depend on the right biochemical and medical questions being asked.

One of the main applications of ^{31}P-NMR with human beings today is the investigation of muscle physiology and pathology, partly because muscles are close to the surface and can easily be measured with surface coils. A very good example of this kind of study is the investigation of a patient with McArdle's syndrome (Ross et al., 1981), which is frequently quoted because of its clarity and which is shown here in Fig. 5.14. The spectra from the patient at rest do not differ from those of a healthy person and show the signals from inorganic phosphate, from phosphocreatine and from ATP. The pH value deduced from the chemical shift is also similar, within the limits of error. As explained above, with a healthy person the level of phosphocreatine decreases under muscular exertion and the P_i level increases. In addition, glucose is released from the muscle glycogen and used for ATP regeneration. Since the oxygen supply is interrupted, only anaerobic glycolysis can occur, the final product of which, lactic acid, accumulates in the muscle and leads to a large decrease in the pH value in the muscle. During the regeneration phase the previous equilibrium state is restored. The patient suffering from McArdle's syndrome lacks muscle phosphorylase, without which no glucose can be produced from glycogen. Hence, no ATP can be regenerated by anaerobic glycolysis and no lactic acid is formed. Therefore, the level of phosphocreatine decreases considerably faster and an intracellular decrease in the pH value due to lactic acid does not occur. On the contrary, a slight increase in the pH value is observed, as would be expected from the creatine kinase reaction (5.2), in which protons are bound as ATP is regenerated. The recovery phase is again not conspicuous. The spectral changes observed can be well explained from the known biochemistry of McArdle's syndrome; their observation can lead to a non-invasive diagnosis of McArdle's syndrome.

However, the practical importance of this investigation is reduced by the fact that although there exist a considerable number of hereditary muscle diseases such as McArdle's syndrome, they very rarely occur. Nevertheless, it is to be hoped that in these cases an unambiguous diagnosis can be reached non-invasively with NMR spectroscopy; furthermore, it provides a method for direct monitoring in the development of new forms of therapy.

Another field where phosphorus resonance may become important is transplantation surgery. Here it can potentially contribute to improving methods of

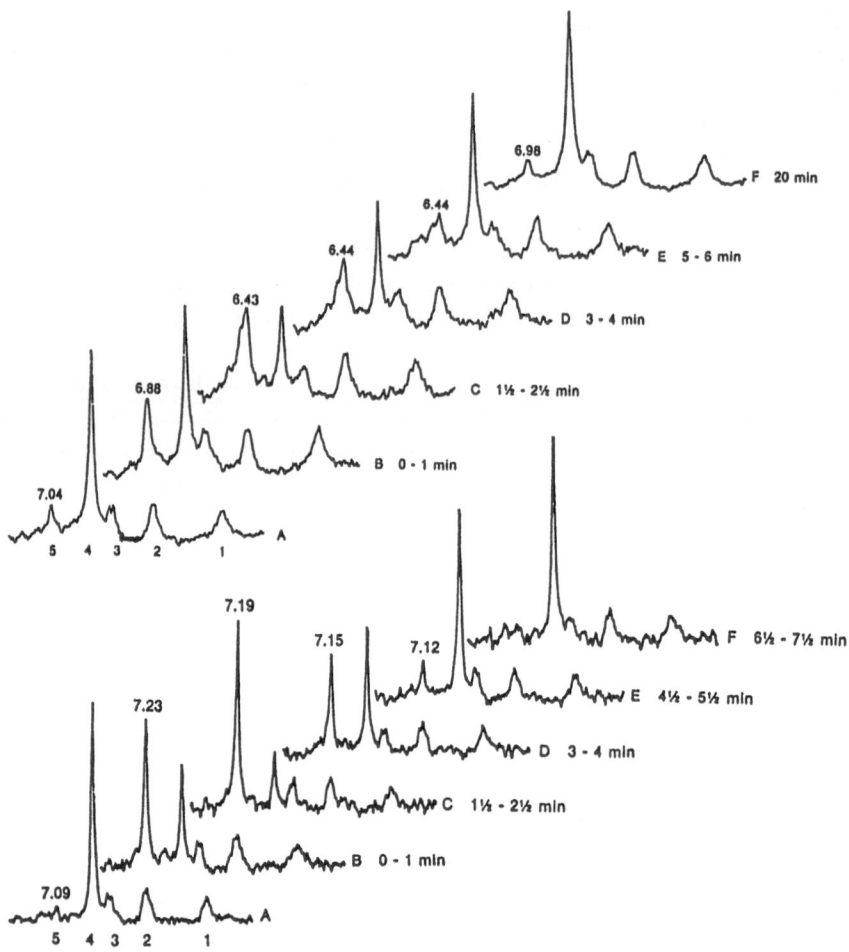

Fig. 5.14. ^{31}P-NMR spectrum in a case of McArdle's syndrome. The ^{31}P-NMR spectra shown were recorded at 1.9 tesla (32.5 MHz) with a surface coil that was fastened to the forearm of a healthy person (above) and of a patient suffering from a McArdle's syndrome (below). At first, a spectrum at rest was taken (A), subsequently, the natural blood supply was interrupted by a blood pressure cuff and simultaneously the muscles were exercised by regularly opening and closing the hand (for 90 s with the healthy person, for 45 s with the patient). After three minutes the normal blood supply was restored. The signals from the inorganic phosphate (5), from the phosphocreatine (4) and from the γ-, α- and β-phosphate groups of the ATP (3, 2 and 1, respectively) are shown in the spectra. The pH values were determined from the chemical shift of the P_i signal (Ross et al., 1981, with permission)

conserving organs, and it can help in assessing the chances of survival of organs that have been or will be transplanted.

A further major field of application of in vivo NMR is the investigation of cancer cells and tumors of various origins. In this context the investigation of the metabolism of the tumor plays an important role: it shows that in exper-

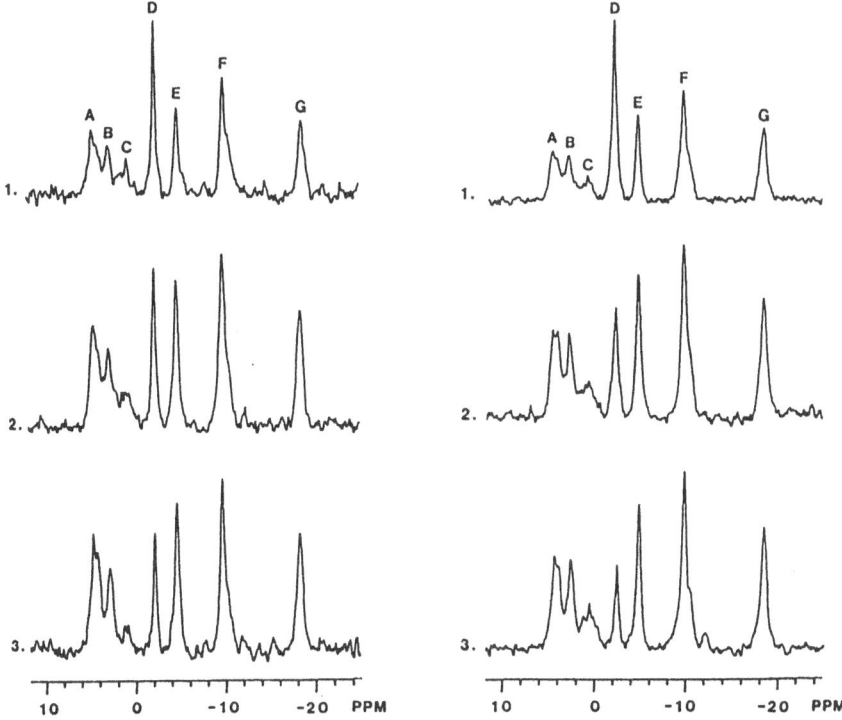

Fig. 5.15. ^{31}P-NMR spectrum of an intracerebral tumor. The phosphorus resonance spectra (8.5 Tesla, 145 MHz) of the left (left) and of the right (right) half of the brains of living mice are shown. The spectra were taken from a healthy mouse (1) and from a mouse with a tumor implanted into the right part of the brain 17 days (2) and 19 days (3) after the implantation. The signals were assigned as follows: (A) phosphomonoester (AMP, sugar phosphate), (B) inorganic phosphate, (C) phosphodiester (glycerophosphorylethanolamine, glycerophosphorylcholine), (D) phosphocreatine, (E) γ-phosphate of the nucleoside triphosphate (ATP) and β-phosphate of the nucleoside diphosphate (ADP), (F)α-phosphate of the nucleoside tri- and diphosphate (ATP,ADP), (G) β-phosphate of the nucleoside triphosphate (ATP) (Ross et al., 1987, with permission)

imental tumors, the concentrations of phosphorus containing metabolites, and the corresponding specific longitudinal and transverse relaxation times, exhibit a characteristic dependence on the type and state of development of the tumor. In Fig. 5.15 ^{31}P-NMR spectra of the brain of a healthy mouse and a mouse with an implanted brain tumor (KHT tumor) are reproduced in different stages of development. In the tissue of the tumor the concentration of phosphocreatine is markedly reduced. In addition, there is a small peak at −12.25 ppm, which may possibly originate from uridine diphosphoglucose (UDPG), an intermediate of glycogen biosynthesis, that cannot be detected in normal brain tissue. The stage of development of a tumor characteristically influences its phosphorus resonance spectrum as well; at a later stage of development the vascularisation may be impaired. The pH value then decreases through the increase of anaerobic glycolysis and can be correlated with the chemical shift of the P_i resonance. Finally, in the late necrotic state energy-providing phosphates can no longer be detected.

The investigation of metabolic processes of phosphorus containing metabolites following different therapies may become an interesting application of ^{31}P-NMR in tumor biology. It could provide new information on the reaction mechanisms; ^{31}P-NMR could become still more important in the future if it became possible to control the efficiency of cancer therapies or the individual reaction to specific forms of therapy by in vivo NMR in human beings, as preliminary results suggest may be feasible.

5.2.4 In Vivo ^{13}C-NMR Spectroscopy

Carbon is one of the elements that occur in almost all molecules of biological interest. The isotope ^{12}C, which exists with the highest natural abundance, is an even-even nucleus, that is a nucleus composed of an even number of protons and an even number of neutrons. The nuclear spin of even-even nuclei in the nuclear ground state is always zero. The only isotope which can be used for NMR is ^{13}C which has a nuclear spin of $\frac{1}{2}$. The natural abundance of this isotope is rather low, approximately 1.1 %. Since the magnetogyric ratio is also not very high, ^{13}C-NMR spectra can only be obtained in an acceptable time from substances that occur in high concentration (with a field strength of 11 tesla and a volume of 2 ml, about 100 mM). However, this limiting value can be significantly increased by an isotope enrichment and by using larger sample volumes. But one has to distinguish here between selective and non-selective enrichment: in the case of a non-selective enrichment the J-coupling of neighboring ^{13}C atoms leads to a splitting of the resonance lines and thus to a reduction of the intensity of the individual lines. Consequently, if the enrichment of the isotopes is done only for a general increase in the intensity, for instance by growing the bacteria on a ^{13}C-enriched nutritious medium, the sensitivity of detection first increases with the enrichment and then decreases again with higher concentrations, owing to the increase in probability that a ^{13}C atom has another ^{13}C neighbor in the molecule. A simple application of the binomial distribution leads to a favorable enrichment of 10 to 20 %. This restriction does not apply, of course, with selective enrichment in a single position in the molecule.

In biomolecules one or several hydrogen atoms are directly bound to most carbon atoms. The resulting line splitting of the ^{13}C resonance lines provides essentially only information on the number of the directly bound protons. Since the line splitting simultaneously leads to a reduction of the intensity, one often forgoes this information and suppresses the line splitting by proton decoupling during data recording. If one irradiates for a sufficient time in the frequency range of proton resonance before recording, polarization can be transferred from the protons to the ^{13}C spins. With small molecules a nuclear Overhauser enhancement up to 2.98 can be obtained, which is a considerable increase of the sensitivity of detection. The actual enhancement of sensitivity obtained, however, depends on many factors that determine the magnitude of the NOEs. Hence, the relative concentrations can no longer be reliably determined by measuring the areas below the resonance lines.

An advantage of ^{13}C-NMR is certainly the large range of chemical shifts of more than 200 ppm. However, with complex systems a strong superposition of resonance lines occurs in spite of the in principle high spectral resolution of ^{13}C-NMR, so that a simplification of the spectra is required similar to ^{1}H-NMR. To accomplish this, most of the schemes described with proton resonance can be transferred to ^{13}C-NMR. However, because of the low sensitivity of ^{13}C-NMR,

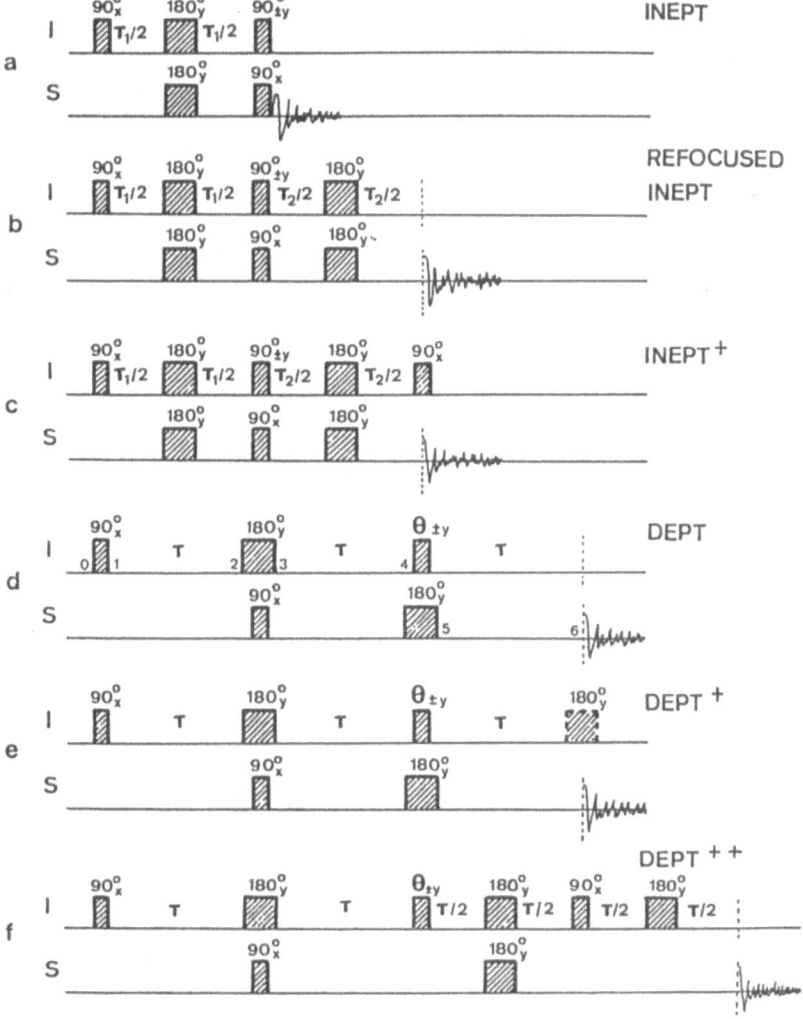

Fig. 5.16. Methods for ^{1}H-^{13}C polarization transfer. The polarization of the I spins (^{1}H) is transferred to the S spins (^{13}C) by the pulse sequences shown (Sørenson and Ernst, 1983, with permission)

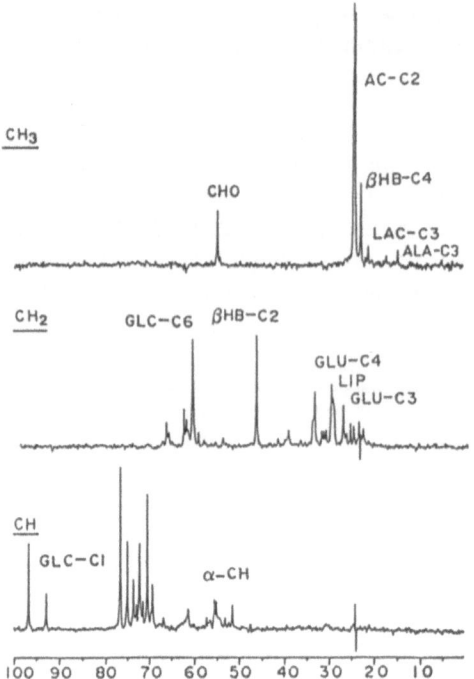

CH₃

CHO
AC-C2
βHB-C4
LAC-C3
ALA-C3

CH₂
GLC-C6 βHB-C2
GLU-C4
LIP
GLU-C3

CH
GLC-CI
α-CH

100 90 80 70 60 50 40 30 20 10

Fig. 5.17. Edited DEPT spectra of a perfused rat liver. The rat liver shown was perfused for one hour before the measurement with a buffer that was enriched with 15 mM [2-^{13}C]acetate. The ^{13}C spectra were recorded at 90.55 MHz corresponding to a proton frequency of 360.13 MHz. The subspectra originate from a suitable combination of the DEPT spectra that were recorded with different flip angles for the protons ($\pi/4$, $\pi/2$ and $3\pi/4$). The following resonances were assigned: The CH₃ groups of choline (CHO), acetate (AC-C2), β-hydroxybutyrate (βHB-C4), lactate (LAC-C3) and alanine (ALA-C3); the CH₂ groups of glucose (GLC-C6), β-hydroxybutyrate (βHB-C2), glutamate (GLU-C3 and GLU-C4) and lipids (LIP); the CαH groups of the amino acids (α-CH); and the C₁H groups of α- and β-glucose (GLC-C1) (Bendall et al., 1985, with permission)

one tries at the same time to increase the sensitivity. Important methods for achieving this are DEPT (distortionless enhancement by polarization transfer) and INEPT (insensitive nuclei enhanced by polarization transfer), as well as variations of these that increase the sensitivity of detection of ^{13}C by polarization transfer from the directly coupled protons (Figs. 5.16 and 5.17). Under practical conditions the theoretically possible gain in sensitivity often cannot be completely reached because the necessary B_1 field homogeneity for these experiments is not present in biological samples with high concentrations of electrolytes.

The scalar coupling of carbon to the directly bound protons is rather large and can ordinarily be well resolved. One ^{13}C nuclear spin leads to a doublet splitting of the corresponding ^{1}H resonance lines of more than 100 Hz, which means that ^{13}C can be indirectly observed by the corresponding proton signal. A very simple method is a modification of the simple spin echo difference method. A proton spin echo ($\frac{\pi}{2}[H, x] - \tau - \pi[H, y] - \tau$) is generated with non-selective pulses where the time τ is chosen in such a way that the multiplet components produced by interaction with the ^{13}C spin are in antiphase ($\tau = J/2$). The ^{1}H-180° pulse leads to a refocusing of the magnetization on the y axis at time 2τ. If one irradiates a ^{13}C-180° pulse simultaneously with the ^{1}H-180° pulse, one observes, analogously to the homonuclear case, a refocusing of the ^{13}C-coupled components on the $-y$ axis; that is the absorption signal becomes negative. If one takes the difference between the spectra with and without the ^{13}C pulse, then only the resonance lines of the ^{13}C-coupled spins remain. Figure 5.18 is an example of such a spin echo difference spectrum where the metabolism of ^{13}C-labelled glucose and the production of ethanol by yeast cells can be observed by using this simple method.

In solution there is an equilibrium between two ring-shaped isomers of the commonly occurring sugar D-glucose, namely the α and β anomers, and an open form, which, however, occurs only in very low concentration. Both anomers can be distinguished in the NMR spectrum by their two separate doublet lines (Fig. 5.18). In the case of anaerobic glycolysis, the yeast cells use glucose as an energy source by converting it via several intermediate steps into ethanol, the resonance lines of which become steadily more intense with time, while after some time signals of glucose can no longer be detected. Although the production of alcohol from glucose is not a new discovery, this simple example demonstrates how with this method the turnover of ^{13}C-labelled substrates can be continuously followed with a good signal-to-noise ratio.

Other methods for detecting ^{13}C with good sensitivity are modifications of the polarization transfer described above, where the polarization is finally retransferred to the protons and thus detected (inverse INEPT, inverse DEPT). A repeatedly occurring problem when detecting via ^{1}H-NMR is that the signal of interest is several orders of magnitude weaker than the water signal if the measurement is not done in D_2O. Although the water signal should theoretically not be visible because water does not contain carbon, an incomplete elimination leads to a distortion of the base line and to superpositions. Nevertheless, a modification

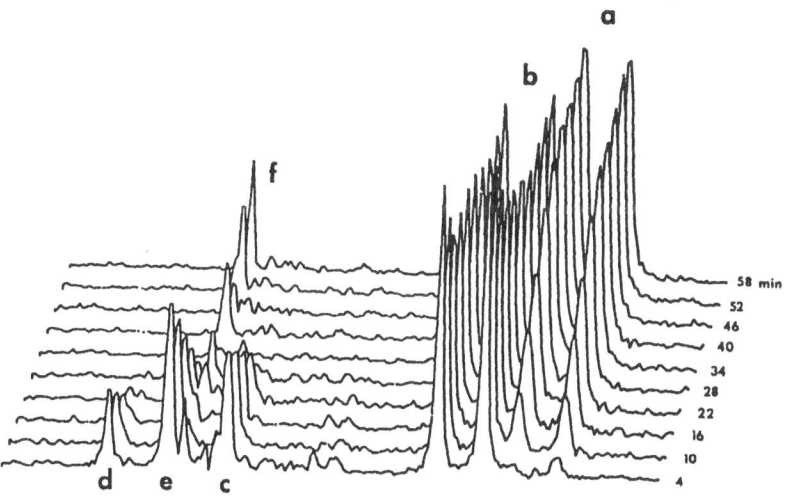

Fig. 5.18. The turnover of [1-^{13}C]glucose by yeast cells. ^{1}H[^{13}C] spin echo difference spectra of a yeast suspension after adding [1-^{13}C]glucose and [2-^{13}C]acetate. The resonance lines can be assigned to [2-^{13}C]ethanol (a), [2-^{13}C]acetate (b), β-[1-^{13}C]glucose (c) and α-[1-^{13}C]glucose (d). Line (e) originates from the superpositions of the signals of the α- and β-anomers of glucose, line (f) is the remaining part of the water line (Foxall et al., 1983, with permission)

of the inverse DEPT experiment allows the elimination of a water signal that is more than 30 000 times more intense.

A disadvantage of all these methods is, however, that the high resolution of the ^{13}C-NMR is lost, that is, their application is only worthwhile with simple spectra as they are obtained by selective isotope labelling of just one substance. If one really wants to obtain the high resolution of the ^{13}C-NMR in a multicomponent system, either ^{13}C must be detected directly or heteronuclear correlated methods must be applied (Sect. 2.3.4) which, however, require markedly longer measuring times.

As with phosphorus resonance, there are today a number of different ^{13}C-NMR investigations of cells, tissues and intact organs. Often ^{13}C enriched substrates are measured, the turnover of which is followed with NMR. The liver

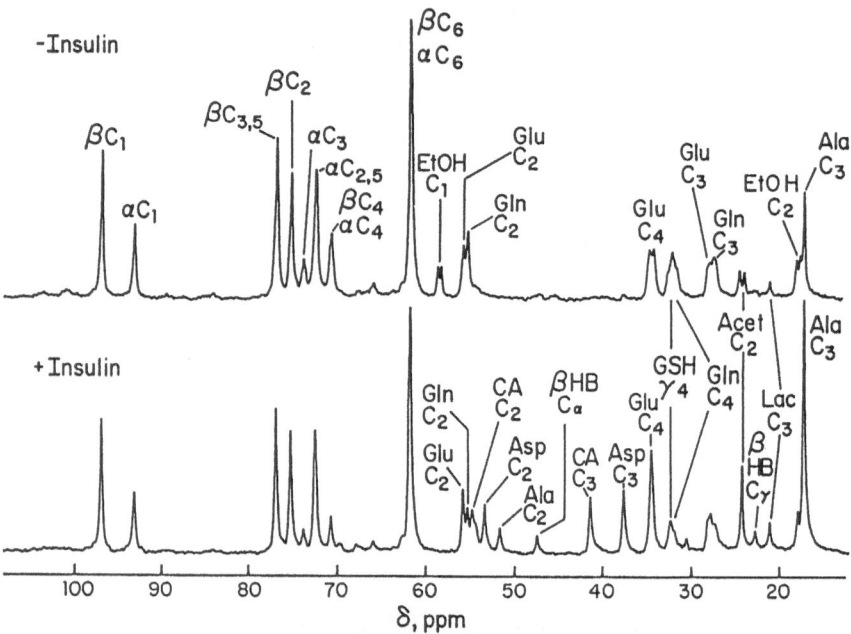

Fig. 5.19. The metabolism of ^{13}C-labelled substrates in the liver. ^{13}C-NMR spectra (90.5 MHz) of the livers of diabetic rats. The livers were perfused with *(upper spectrum)* and without addition of insulin *(lower spectrum)* for 170 minutes with 10 mM [3-^{13}C]alanine and, respectively, with 7.3 mM [1,2-^{13}C]ethanol *(upper spectrum)* and [2-^{13}C]ethanol *(lower spectrum)*. The starting spectra before adding the enriched substrates were subtracted from both spectra. The signals observed originate therefore predominantly from decay products of the ^{13}C-labelled substrates. The ^{13}C-NMR lines were assigned to the C-1 and C-2 of ethanol (EtOH), the C-2 and C-3 of alanine (Ala), the C-2 of acetate (Acet), the C-3 of lactate (Lac), the C-2, C-3 and C-4 of glutamine (Gln), the C-2, C-3 and C-4 of glutamate (Glu), the C-4 of the glutamyl residue in glutathione (GSH), the C-1, C-2, C-3, C-4, C-5 and C-6 of the α- and β-anomers of glucose (α, β), the C-2 and C-3 of β-hydroxybutyric acid (βHB), the C-2 and C-3 of aspartate (Asp) and the C-2 and C-3 of N-carbamyl-aspartate (CA). The part of the spectrum in which the resonance lines of the carboxyl groups occur is not shown (Cohen, 1987, with permission)

as a metabolically active organ is of course particularly interesting for experiments that study the incorporation of ^{13}C atoms into metabolites. If rat livers are perfused with solutions containing ^{13}C-labelled substrates, the flow of these substrates through different metabolic pathways can be followed with a time resolution of about 10 min depending on external conditions (hormonal stimulation, composition of the perfusion solution) and anamnesis (well-fed, after a longer fasting period, diabetic). Such experiments were already performed earlier using radioactive ^{14}C-labelled substrates, the incorporation of which into reaction products was then analysed. However, using NMR such experiments can be performed noninvasively and the concentrations of many different metabolites can be determined simultaneously and continuously. Figure 5.19 shows as an example ^{13}C-NMR spectra obtained after 170 minutes of perfusion of livers of diabetic rats with [3-^{13}C]alanine and using [2-^{13}C]ethanol with insulin and [1,2-^{13}C]ethanol without insulin. These spectra reveal directly into which new molecules the ^{13}C fragments are incorporated during the metabolic processing of alanine and ethanol. Of the different possibilities to metabolize [3-^{13}C]alanine via [3-^{13}C]pyruvate, one pathway leads to gluconeogenesis, that is, to a new synthesis of glucose, and another to the resynthesis of the starting products (Fig. 5.20). In the course of the tricarboxylic acid cycle the ^{13}C atom of [3-^{13}C]alanine is incorporated into position 2 of phospho*enol*pyruvate and appears after several completed cycles also in position 1 of phospho*enol*pyruvate. Accordingly, the distribution of the ^{13}C atoms in the reaction products glucose and alanine is changed also. An analysis of this distribution permits to investigate the turnover

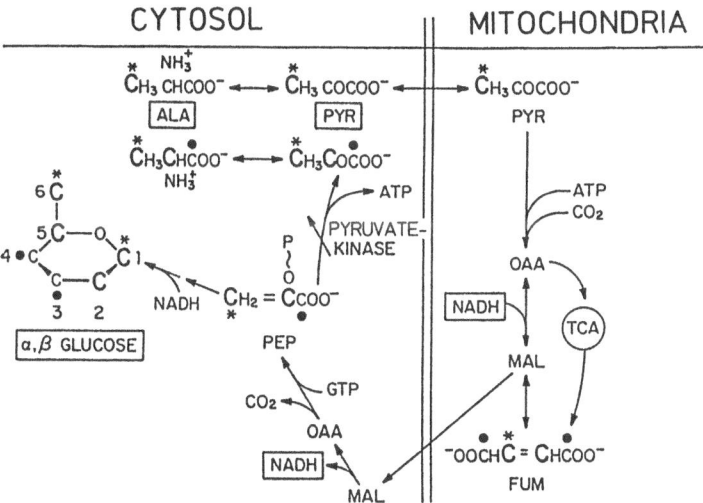

Fig. 5.20. Simplified model of gluconeogenesis. The original ^{13}C labelling of alanine (ALA) is marked by large letters and an asterisk and the black dots denote the C atoms which contain ^{13}C after repeated passages through the tricarboxylic acid cycle. Abbreviations: tricarboxylic acid cycle (TCA), oxalacetate (OAA), pyruvate (PYR), malate (MAL), fumarate (FUM), phospho*enol*pyruvate (PEP) (Cohen, 1987, with permission)

via the individual metabolic pathways. From this experiment one finds that the livers of rats that had fasted for 24 hours before the experiment resynthesize about the same amount of alanine to its starting product (futile cycle) as they use for the new synthesis of glucose.

Information about whether one and the same molecule has more than one label is difficult to obtain with classical methods. With NMR this is very simple

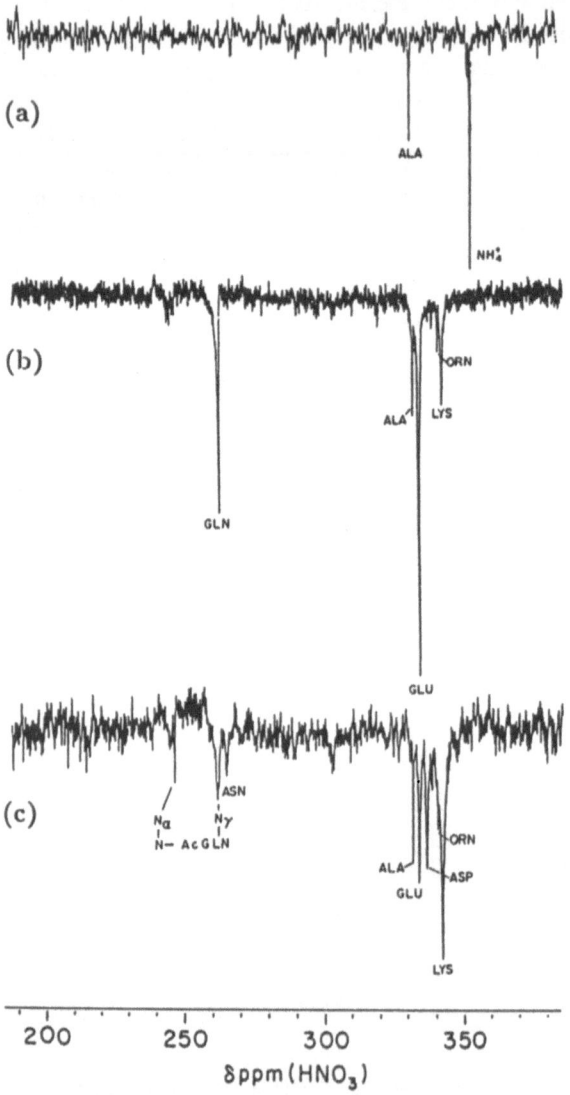

Fig. 5.21a–c. ^{15}N-NMR spetra of *B. lactofermentum*. The broad band decoupled spectra were recorded at 30.4 MHz (proton resonance frequency 300 MHz). Spectrum of the cell-free medium in which the cells were bred for 35 hours (a), and the spectra of the bacteria after growing for 17 hours (b) and for 35 hours (c) (Haran et al., 1983, with permission)

because it causes a splitting of the lines by scalar coupling. This method was additionally applied in the experiment described above (Fig. 5.19) for studying the decomposition of ethanol. Doubly labelled ethanol [1,2-^{13}C]ethanol can be seen as a doublet resonance in the upper spectrum of Fig. 5.19. Both atoms are incorporated together into acetate and glutamate, as can be easily seen from their doublet structure. Many possible metabolic pathways can be investigated in detail in a similar manner; the imagination of the experimentalist is here only limited by the sensitivity of detection.

5.2.5 In Vivo ^{15}N-NMR Spectroscopy

The nitrogen isotope with the highest natural abundance is ^{14}N, which has a nuclear spin $I = 1$. Owing to its quadrupole moment the corresponding resonance lines are mostly so strongly broadened that this isotope does not play a practical role in biological NMR. This is entirely different with the nitrogen isotope ^{15}N, which is very suitable for NMR with its nuclear spin $I = \frac{1}{2}$. Since its natural abundance is only 0.37 %, it is even more suitable for isotope labelling experiments than ^{13}C whose natural abundance is 1.1 %, because the perturbing signal of the background is correspondingly weaker.

When broadband decoupling is applied to the protons, the negative magnetogyric ratio of ^{15}N can lead to the unfavorable situation that the corresponding nuclear Overhauser effect causes the signal to vanish completely (see Sect. 2.6.6). All methods already described with ^{13}C-NMR are also suitable for the more sensitive detection of the ^{15}N signal. Since the range of chemical shift is very large, well resolved spectra can be expected.

Figure 5.21 shows as an example the ^{15}N-NMR spectrum of *Brevibacterium lactofermentum* that was grown on a medium containing ^{15}NH$_4$Cl as the source of nitrogen. The incorporation of the labelled nitrogen into the cells can be easily studied with this method. After some time, the ^{15}N labelled amino acid alanine (Ala) can be detected outside the cells, which means it is passed on to the medium by the bacteria, while the other amino acids remain completely within the cells.

5.2.6 In Vivo ^{19}F-NMR Spectroscopy

Although the NMR active isotope fluorine ^{19}F occurs with a natural abundance of 100 %, it can serve for labelling substances as well as ^{15}N because most natural systems in the soluble phase contain practically no fluorine. A great advantage of this $I = \frac{1}{2}$ nucleus is, besides its large range of chemical shift, its high NMR sensitivity which is similar to that of protons. Substances containing fluorine are frequently used as drugs and the study of their metabolism is a natural field of application of ^{19}F-NMR.

The anaesthetics halothane and isoflurane contain fluorine so that their enrichment and metabolism in tissue can be studied. Figure 5.22 shows the decrease of isoflurane in brain tissue after an isoflurane anaesthesia. Halothane remains

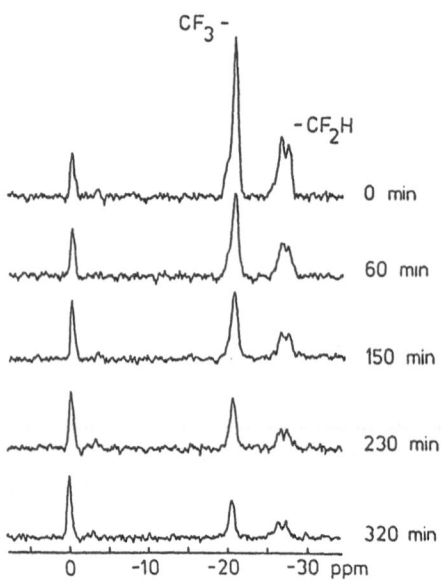

Fig. 5.22. ^{19}F-NMR of an anaesthetic in brain tissue. ^{19}F-NMR spectra (75.5 MHz) of isoflurane ($CF_3CHClOCF_2H$) in the brain of a living rabbit that was anaesthesized with isoflurane before the measurement. The NMR measurement shows the decay of the anaesthetic in the tissue. The signal at 0 ppm belongs to the external reference dibromotetrafluorethane. The measurement was performed with a surface coil (Wyrwicz et al., 1987, with permission)

stable in most tissues but is partially transformed into toxic substances by the liver. Four different fluorine-containing halothane decomposition products could be detected in the livers of rats that were anaesthesized with halothane.

Similar experiments can also be performed with antimetabolites for treating cancer such as difluoromethylornithine (DFMO), an inhibitor of ornithinedecarboxylase, which several tumors tend to absorb selectively. Besides its importance for fundamental research, this kind of investigation could also be of direct clinical relevance for the spectroscopy of humans if one could deduce directly from the fluorine spectra of a drug whether or not the concentration in the target, for instance in a tumor, is sufficiently high.

5.2.7 In Vivo Spectroscopy with Other Nuclei

Of course, in vivo NMR with all other nuclei with a non-zero nuclear spin is possible. The quadrupolar nuclei [23]Na and [39]K are particularly suitable because they occur in biological systems in high concentration. Deuterium, which has only a low natural abundance, can in principle be used for labelling molecules. Here, however, it is better to detect the protons and to follow the label from the disappearing of the ^{1}H signal.

The resonance lines of the quadrupolar nuclei are in general so broad that the differences in the chemical shifts frequently do not suffice to separate the resonance lines of different molecules under in vivo NMR conditions. Furthermore, sodium and potassium ions are not covalently bound in biological molecules and therefore their NMR resonance frequencies are not characteristically shifted. Their interaction with macromolecules in the cell causes only a decrease in the relaxation times. Occasionally, multiexponential decay curves are observed if the ions are in different chemical surroundings.

References

Chapter 1

Arnold, J.T., Dharmatti, S.S. and Packard, M.E. (1951) J. Chem. Phys. **19**, 507
Bloch, F. (1946) Phys. Rev. **70**, 460
Carr, H.Y. and Purcell, E.M. (1954) Phys. Rev. **94**, 630
Ernst, R.R. and Anderson, W.A. (1966) Rev. Sci. Instrum. **37**, 93
Jeener, J. (1971) presented at the Ampère Summer School in Basko Polje
Knight, W.D. (1949) Phys. Rev. **76**, 1259
Lauterbur, P.C. (1973) Nature **242**, 190
Mansfield, P., Grannell, P.K., Garroway, A.N. and Stalker, D.C. (1973) Proc. Specialized Colloque Ampère, Krakau
Meiboom, S. and Gill, D. (1958) Rev. Sci. Instrum. **29**, 688
Purcell, E.M., Torrey, H.C. and Pound, R.V. (1946) Phys. Rev. **69**, 37

Further Reading

Abragam, A. (1961) *The Principles of Nuclear Magnetism* (Oxford University Press, Oxford)
Ernst, R.R., Bodenhausen, G. and Wokaun, A. (1990) *Principles of Nuclear Magnetic Resonance in One and Two Dimensions* (Oxford University Press, Oxford)
Poole, C.P. (1990) *Theory of NMR* (Wiley, London)
Shaw, D. (1976) *Fourier Transform N.M.R. Spectroscopy* (Elsevier, Amsterdam)
Slichter, C.P. (1990) *Principles of Magnetic Resonance* (Springer, Berlin, Heidelberg, New York)

Chapter 2

McConnell, H.M. (1958) J. Chem. Phys. **28**, 430
Overhauser, A.W. (1953) Phys. Rev. **92**, 411
Rösch, P., Kalbitzer, H.R., Schmidt-Aderjan, U. and Hengstenberg, W. (1981a) Biochemistry **20**, 1599
Rösch, P., Goody, R.S., Kalbitzer, H.R. and Zimmermann, H. (1981b) Arch. Biochem. Biophys. **211**, 622
Sontheimer, G.M., Kuhn, W. and Kalbitzer, H.R. (1986) Biochem. Biophys. Res. Comm. **134**, 1379

Further Reading

Brey, W.S. (ed.) (1988) *Pulse Methods in 1D and 2D Liquid-Phase NMR* (Academic, San Diego)
Carrington, A. and McLachlan, A.D. (1969) *Introduction to Magnetic Resonance with Applications to Chemistry and Chemical Physics* (Harper and Row, New York)
Chandrakumar, N. and Subramanian, S. (1987) *Modern Techniques in High-Resolution FT-NMR* (Springer, New York, Berlin, Heidelberg)
Croasmun, W.R. and Carlson, R.M.K. (1987) *Two-Dimensional NMR Spectroscopy, Applications for Chemists and Biochemists* (VCH, Weinheim)

Günther, H. (1980) *NMR Spectroscopy* (Wiley, New York)

Harris, R.K. (1983) *Nuclear Magnetic Resonance Spectroscopy* (Pitman, Marshfield)

Martin, M.L., Martin, G.J. and Delpuech, J.-J. (1980) *Practical NMR Spectroscopy* (Heyden, London)

Sanders, J.K.M. and Hunter, B.K. (1987) *Modern NMR Spectroscopy, A Guide for Chemists* (Oxford University Press, Oxford)

Chapter 3

Abraham, R.J., Fell, S.C.M. and Smith, K.M. (1977) Org. Magn. Res. **9**, 367

Berden, J.A., Cullis, P.R., Hoult, D.I., McLaughlin, A.C., Radda, G.K. and Richards, R.E. (1974) FEBS Lett. **46**, 55

Berden, J.A., Barker, R.W. and Radda, G.K. (1975) Biochem. Biophys. Acta **375**, 186

Bertini, I., Luchinat, C. and Messori, L. (1987) in *Metal Ions in Biological Systems*, Vol. 21, edited by H. Siegel (Marcel Dekker, Basel)

Billeter, M., Braun, W. and Wüthrich, K. (1982) J. Mol. Biol. **155**, 321

Bleaney, B., Dobson, C.M., Levine, B.A., Martin, R.B., Williams, R.J.P. and Xavier, A.V. (1972) J.C.S. Chem. Comm., 791

Boelens, R., Scheek, R.M., van Boom, J.H. and Kaptein, R. (1987) J. Mol. Biol. **193**, 213

Braun, W., Bösch, C., Brown, L.R., Gō, N. and Wüthrich, K. (1981) Biochem. Biophys. Acta **667**, 377

Braun, W. and Gō, N. (1985) Mol. Biol. **186**, 611

Bretscher, M.S. (1985) Sci. Am. **253** (4) 86

Bundi A. and Wüthrich, K. (1979) Biopolymers **18**, 285

Dabrowski, J., Dabrowski, U., Hanfland, P., Kordowicz, M. and Hull, W.E. (1986) Magn. Res. Chem. **24**, 729

DeMarco, A., Llinás, M. and Wüthrich, K. (1978) Biopolymers **17**, 637

Dickerson, R.E. (1983) Sci. Am. **249** (6) 100

Dubs, A., Wagner, G. and Wüthrich, K. (1979) Biochem. Biophys. Acta **577**, 177

El-Kabbani, O.A.L., Waygood, E.B. and Delbaere, L.T.J. (1987) J. Biol. Chem. **262**, 12926

Frey, M.H., Wagner, G., Vašák, M., Sørensen, O.W., Neuhaus, D., Wörgötter, E., Kägi, J.H.R., Ernst, R.R. and Wüthrich, K. (1985) J. Am. Chem. Soc. **107**, 6847

Frey, M.H., Leupin, W., Sørensen, O.W., Denny, W.A., Ernst, R.R. and Wüthrich, K. (1985) Biopolymers **24**, 2371

Furey, W.F., Robbins, A.H., Clancy, L.L., Winge, D.R., Wang, B.C. and Stout, C.D. (1986) Science **231**, 704

Ghélis, C. and Yon, J. (1982) *Protein Folding* (Academic, New York)

Glaser, S. and Kalbitzer, H.R. (1986) J. Mag. Res. **68**, 350

Glaser, S. (1987) Doctoral Thesis, University of Heidelberg

Gross, K.-H. and Kalbitzer, H.R. (1988) J. Magn. Res. **76**, 87

Haigh, C.W. and Mallion, R.B. (1972) Org. Magn. Res. **4**, 203

Havel, T.F., Kuntz, I.D. and Crippen, G.M. (1983) J. Theor. Biol. **104**, 359

Havel, T.F. and Wüthrich, K. (1985) J. Mol. Biol. **182**, 281

Heerschnap, A., Haasnot, C.A.G. and Hilbers, C.W. (1983) Nucl. Acid Res. **11**, 4501

Hyde, E.I. and Reid, B.R. (1985) Biochemistry **24**, 4315

IUPAC-IUB (1970) J. Biol. Chem. **245**, 6489

Johnson, C.E. Jr. and Bovey, F.A. (1958) J. Chem. Phys. **29**, 1012

Kalbitzer, H.R., Deutscher, J., Hengstenberg, W. and Rösch, P. (1981) Biochemistry **21**, 6178

Kalbitzer, H.R., Marquetant, R., Rösch, P. and Schirmer, R.H. (1982a) Eur. J. Biochem. **126**, 531

Kalbitzer, H.R., Hengstenberg, W., Rösch, P., Muss, P., Bernsmann, P., Engelmann, R., Dörschug, M. and Deutscher, J. (1982b) Biochemistry **21**, 2879

Kaptein, R., Zuiderweg, E.R.P., Scheek, R.M., Boelens, R. and van Gunsteren, W.F. (1985) J. Mol. Biol. **182**, 179

Kim, S.H., Suddath, F.L., Quigley, G.J., McPherson, A., Sussman, J.L., Wang, A.H.J., Seeman, N.C. and Rich, A. (1974) Science **185**, 435

Klevit, R.E. and Waygood, E.B. (1986) Biochemistry **25**, 7774

Kline, A.D., Braun, W. and Wüthrich, K. (1986) J. Mol. Biol. **189**, 377

Kumar, A., Wagner, G., Ernst, R.R. and Wüthrich, K. (1981) J. Am. Chem. Soc. **103**, 3654

Lee, L. and Sykes, B.D. (1983) Biochemistry **22**, 4366

Lehninger, A.L. (1983) *Biochemie* (VCH, Weinheim)

Meier, B.U., Bodenhausen, G. and Ernst, R.R. (1984) J. Magn. Res. **60**, 161

Morishima, I., Shiro, Y. and Nakajiama, K. (1986) Biochemistry **25**, 3576

Opella, S.J., Stewart, P.L. and Valentine, K.G. (1987) Quart. Rev. Biophys. **19**, 7

Pauling, L. (1936) J. Chem. Phys. **4**, 673

Perkins, S.J. and Dwek, R.A. (1980) Biochemistry **19**, 245

Pflugrath, J.W., Wiegand, G., Huber, R. and Vértesy, L. (1986) J. Mol. Biol. **189**, 383

Pople, J.A. (1956) J. Chem. Phys. **24**, 1111

Robbins, A.H., McRoe, D.E., Williamson, M., Collet, S.A., Xuong, N.H., Furey, W.F., Wang, B.C. and Stout, C.D. (1990) J. Mol. Biol. submitted

Schulz, G.E. and Schirmer, R.H. (1979) *Principles of Protein Structure* (Springer, Berlin, Heidelberg, New York)

Seelig, J. and Seelig, A. (1980) Quart. Rev. Biophys. **13**, 19

Tüchsen, E. and Woodward, C. (1985) J. Mol. Biol. **185**, 421

Wagner, G. (1983) Quart. Rev. Biophys. **16**, 1

Wagner, G., Neuhaus, D., Wörgötter, E., Vašák, M., Kägi, J.H.R. and Wüthrich, K. (1986) J. Mol. Biol. **187**, 131

Waugh, J.S., Huber, L.M. and Häberlen, U. (1968) Phys. Rev. Lett. **20**, 180

Wüthrich, K. (1976) *NMR in Biological Research: Peptides and Proteins* (North-Holland, Amsterdam)

Wüthrich, K. and Wagner, G. (1978) TIBS 3, 227

Wüthrich, K. (1986) *NMR of Proteins and Nucleic Acids* (Wiley, New York)

Further Reading

Jardetzky, O. and Roberts, G.C.K. (1981) *NMR in Molecular Biology* (Academic, New York)

Neuhaus, D. and Williamson, M.P. (1989) *The Nuclear Overhauser Effect in Structural and Conformational Analysis* (VCH, Weinheim)

Wüthrich, K. (1976) *NMR in Biological Research: Peptides and Proteins* (North-Holland, Amsterdam)

Wüthrich, K. (1986) *NMR of Proteins and Nucleic Acids* (Wiley, New York)

Chapter 4

Deimling, M., Müller, E., Lenz, G., Barth, K., Fritschy, P., Seiderer, M. and Reinhardt, E.R. (1986) Diagn. Imag. Clinic. Med. **55**, 37

Dixon, W.T. (1984) Radiology **153**, 189

Edelstein, W.A., Hutchison, J.M.S., Johnson, G. and Redpath, T. (1980) Phys. Med. Biol. **25**, 751

Haase, A., Frahm, J., Hänicke, W. and Matthaei, D. (1985) Phys. Med. Biol. **30**, 341

Haase, A. and Frahm, J. (1985) J. Magn. Reson. **64**, 94

Haase, A., Frahm. J., Matthaei, D., Hänicke, W. and Merboldt, K.-D. (1986) J. Mag. Reson. **67**, 258

Hahn, E.L. (1950) Phys. Rev. **80**, 580

Hennig, J., Nauerth, A., Friedburg, H. and Ratzel, D. (1984) Radiologe **24**, 579

Hennig, J. and Friedburg, H. (1986) Magn. Reson. Med. **3**, 844

Hennig, J., Friedburg, H. and Ströbel, B. (1986) J. Comp. Ass. Tomography **10**, 375

Hennig, J., Mueri, M., Brunner, P. and Friedburg, H. (1988) Radiology **166**, 237

Johnson, G., Hutchison, J.M.S., Redpath, T.W. and Eastwood, L.M. (1983) J. Magn. Reson. **54**, 374

Kuhn, W. (1990) Ang. Chem. **102**, 1

Kumar, A., Welti, D. and Ernst, R.R. (1975) J. Magn. Reson. **18**, 69

Laub, G. and Kaiser, W. (1988) J. Comp. Ass. Tomography **12**, 377

Laukien, F.H. (1984) B.Sc. Thesis, Massachusetts Institute of Technology, Cambridge, MA

Lauterbur, P. (1973) Nature **242**, 190

Mansfield, P. (1977) J. Phys. C 10, L55

Moran, P.R. (1982) Magn. Reson. Imag. **1**, 197

Müller, E., Deimling, M. and Reinhardt, E.R. (1986) Magn. Reson. Med. **3**, 331

Oppelt, A., Graumann, R., Barfuss, H., Fischer, H., Hartl, W. and Schajor, W. (1986) Electromedica **54**, 15

Petersen, S.B., Muller, R.N. and Rinck, P.A. (1985) *An Introduction to Biomedical Nuclear Magnetic Resonance* (Thieme, New York)

Further Reading

Berquist, T.H., Ehman, R.L. and Richardson, M.L. (1987) *Magnetic Resonance of the Musculosketal System* (Raven, New York)

Bradley, W.G. and Bydder (1990) *MRI Atlas of the Brain* (Deutscher Ärzteverlag, Köln)

Brant-Zawadzky, M. and Norman, D. (1986) *Magnetic Resonance Imaging of the Central Nervous System* (Raven, New York)

Gerhardt, P. and Frommhold, W. (1988) *Atlas of Anatomic Correlations in CT and MRI* (Thieme, Stuttgart)

Higgins, C.B. and Hricak, H. (eds.) (1987) *Magnetic Resonance Imaging of the Body* (Raven, New York)

Huk, W.J., Gademann, G.F. and Friedmann, G. (1989) *Magnetic Resonance Imaging of Central Nervous System Diseases* (Springer, Berlin, Heidelberg, New York)

Kazner, E., Wende, S., Grumme, T., Stochdorph, O., Felix, R. and Claussen, C. (1989) *Computed Tomography and Magnetic Tomography of Intracranial Tumors. A Clinical Perspective* (Springer, Berlin, Heidelberg, New York)

Mansfield, P. and Morris, P.G. (1982) *NMR Imaging in Biomedicine* (Academic, New York)

Matwiyoff, N.A. (1989) *Magnetic Resonance Workbook* (Raven, New York)

Morris, P.G. (1986) *Nuclear Magnetic Resonance Imaging in Medicine and Biology* (Oxford University Press, Oxford)

Partain, C.L., James, A.E., Rollo, F.D. and Price, R.R. (1983) *Nuclear Magnetic Resonance Imaging* (Saunders, Philadelphia)

Partain, C.L., Price, R.R., Patton, J.A., Kulkarni, M.V. and James, A.E. Jr. (1988) *Magnetic Resonance Imaging* (Saunders, Philadelphia)

Petersen, S.B., Muller, R.N. and Rinck, P.A. (1985) *An Introduction to Biomedical Nuclear Magnetic Resonance* (Thieme, New York)

Schulthess, G.K. von (1989) *Morphology and Function in MRI, Cardiovascular and Renal Systems* (Springer, Berlin, Heidelberg, New York)

Stark, D.D. and Bradley, W.G. (1988) *Magnetic Resonance Imaging* (Mosby, St. Louis)

Wehrli, F.W., Shaw, D. and Kneeland J.B. (1988) *Biomedical Magnetic Resonance Imaging: Principles, Methodology, and Applications* (VCH, Weinheim)

Young, S.W. (1988) *Magnetic Resonance Imaging. Basic Principles* (Raven, New York)

Chapter 5

Arnold, D.L., Matthews, P.M. and Radda, G.K. (1984) Magn. Reson. Med. **1**, 307

Bendall, M.R. and Pegg, D.T. (1985) Magn. Reson. Med. **2**, 91

Bendall, M.R., den Hollander, J.A., Arias-Mendoza, F., Rothman, D.L., Behar, K.L. and Shulman, R.G. (1985) Magn. Reson. Med. **2**, 56

Brown, F.F., Campbell, I.D., Kuchel, P.W. and Rabenstein, D.C. (1977) FEBS Lett. **82**, 12

Cohen, S. (1987) Biochemistry **26**, 573

Degani, H., Laughlin, M., Campbell, S. and Shulman, R.G. (1985) Biochemistry **24**, 5510

Foxall, D.L., Cohen, J.S. and Tschudin, R.G. (1983) J. Magn. Reson. **51**, 330

Gadian, D.G., Proctor, E., Williams, S.R., Cady, E.B. and Gardiner, R.M. (1986) Magn. Reson. Med. **3**, 150

Haase, A., Hänicke, W. and Frahm, J. (1984) J. Magn. Reson. **56**, 401

Haase, A. (1986) Magn. Reson. Med. **3**, 963

Haran, N., Kahana, Z.E. and Lapidot, A. (1983) J. Biol. Chem. **258**, 12929

Lyon, R.C., Faustino, P.J. and Cohen, J.S. (1986) Magn. Reson. Med. **3**, 663

Maudsley, A.A. and Hilal, S.K. (1985) Magn. Reson. Med. **2**, 218

Ross, B.D., Radda, G.K., Gadian, D.G., Rocker, G., Esiri, M. and Falconer-Smith, J. (1981) New Engl. J. Med **304**, 1338

Ross, B.D., Higgins, R.J., Conley, F.K. and True, N.S. (1987) Magn. Reson. Med. **4**, 323

Rothman, D.L., Arias-Mendoza, F., Shulman, G.I. and Shulman, R.G. (1984) J. Magn. Reson. **60**, 430

Sørensen, O.W. and Ernst, R.R. (1983) J. Magn. Reson. **51**, 477

Taylor, D.J., Styles, P., Matthews, P.M., Arnold, D.A., Gadian, D.G., Bore, P. and Radda, G.K. (1986) Magn. Reson. Med. **3**, 44

Ugurbil, K., Rottenberg, H., Glynn, P. and Shulman, R.G. (1982) Biochemistry **21**, 1068

Wyrwicz, A.M., Conboy, C.B., Ryback, K.R., Nichols, B.G. and Eisele, P. (1987) Biochem. Biophys. Acta **927**, 86

Subject Index

Ureter 161
Urine 161

Valine *76f*, 83, 105f
Vascular bundles 170
Velocity profile, parabolic 162f
Vesicle, sarcoplasmic 133f
Viscosity 92, 165
VO^{2+} ion 97
Voltage 10, 35, 177f
Volume element *147*, 168f, 179
Voxel, *see* volume element

WAHUHA 119
WALTZ 48
Water accessibility 89
Water signal 39, *50*, 159f, 183
Whole body tomograph 26, 34, 135f

X-ray computer tomography 135, 149, 158
X-ray structure analysis 2, 89, 102, 115f

Zeeman level 17, 30